FORUM ON NEUROSCIENCE AND NERVOUS SYSTEM DISORDERS

AUTISM AND THE ENVIRONMENT

Challenges and Opportunities for Research

WORKSHOP PROCEEDINGS

Board on Health Sciences Policy

INSTITUTE OF MEDICINE
OF THE NATIONAL ACADEMIES

THE NATIONAL ACADEMIES PRESS
Washington, D.C.
www.nap.edu

THE NATIONAL ACADEMIES PRESS • 500 Fifth Street, N.W. • Washington, DC 20001

NOTICE: The project that is the subject of this report was approved by the Governing Board of the National Research Council, whose members are drawn from the councils of the National Academy of Sciences, the National Academy of Engineering, and the Institute of Medicine. The members of the committee responsible for the report were chosen for their special competences and with regard for appropriate balance.

This project was supported by contracts between the National Academy of Sciences and the Alzheimer's Association; Amgen Inc.; AstraZeneca Pharmaceuticals, Inc.; the Centers for Disease Control and Prevention; the Department of Health and Human Services' National Institutes of Health (NIH, Contract No. N01-OD-4-213) through the National Institute on Alcohol Abuse and Alcoholism, the National Institute on Aging, the National Institute on Drug Abuse, the National Eye Institute, the NIH Blueprint for Neuroscience Research, the National Institute of Mental Health, and the National Institute of Neurological Disorders and Stroke; Eli Lily and Company; GE Healthcare, Inc.; GlaxoSmithKline, Inc.; Johnson & Johnson Pharmaceutical Research and Development, Inc.; Merck Research Laboratories, Inc.; the National Multiple Sclerosis Society; the National Science Foundation (Contract No. OIA-0647541); Pfizer Global Research and Development, Inc.; and the Society for Neuroscience. The views presented in this publication are those of the editors and attributing authors and do not necessarily reflect the view of the organizations or agencies that provided support for this project.

International Standard Book Number-13: 978-0-309-10881-2
International Standard Book Number-10: 0-309-10881-0

Additional copies of this report are available from the National Academies Press, 500 Fifth Street, N.W., Lockbox 285, Washington, DC 20055; (800) 624-6242 or (202) 334-3313 (in the Washington metropolitan area); Internet, http://www.nap.edu.

For more information about the Institute of Medicine, visit the IOM home page at: **www.iom.edu.**

Printed in the United States of America.

Suggested citation: Institute of Medicine (IOM). 2008. *Autism and the environment: Challenges and opportunities for research. Workshop proceedings.* Washington, DC: National Academies Press.

"Knowing is not enough; we must apply.
Willing is not enough; we must do."
—Goethe

INSTITUTE OF MEDICINE
OF THE NATIONAL ACADEMIES

Advising the Nation. Improving Health.

THE NATIONAL ACADEMIES

Advisers to the Nation on Science, Engineering, and Medicine

The **National Academy of Sciences** is a private, nonprofit, self-perpetuating society of distinguished scholars engaged in scientific and engineering research, dedicated to the furtherance of science and technology and to their use for the general welfare. Upon the authority of the charter granted to it by the Congress in 1863, the Academy has a mandate that requires it to advise the federal government on scientific and technical matters. Dr. Ralph J. Cicerone is president of the National Academy of Sciences.

The **National Academy of Engineering** was established in 1964, under the charter of the National Academy of Sciences, as a parallel organization of outstanding engineers. It is autonomous in its administration and in the selection of its members, sharing with the National Academy of Sciences the responsibility for advising the federal government. The National Academy of Engineering also sponsors engineering programs aimed at meeting national needs, encourages education and research, and recognizes the superior achievements of engineers. Dr. Charles M. Vest is president of the National Academy of Engineering.

The **Institute of Medicine** was established in 1970 by the National Academy of Sciences to secure the services of eminent members of appropriate professions in the examination of policy matters pertaining to the health of the public. The Institute acts under the responsibility given to the National Academy of Sciences by its congressional charter to be an adviser to the federal government and, upon its own initiative, to identify issues of medical care, research, and education. Dr. Harvey V. Fineberg is president of the Institute of Medicine.

The **National Research Council** was organized by the National Academy of Sciences in 1916 to associate the broad community of science and technology with the Academy's purposes of furthering knowledge and advising the federal government. Functioning in accordance with general policies determined by the Academy, the Council has become the principal operating agency of both the National Academy of Sciences and the National Academy of Engineering in providing services to the government, the public, and the scientific and engineering communities. The Council is administered jointly by both Academies and the Institute of Medicine. Dr. Ralph J. Cicerone and Dr. Charles M. Vest are chair and vice chair, respectively, of the National Research Council.

www.national-academies.org

WORKSHOP ON AUTISM AND THE ENVIRONMENT CHALLENGES AND OPPORTUNITIES FOR RESEARCH PLANNING COMMITTEE*

ALAN LESHNER (*Chair*), American Association for the Advancement of Science, Washington, D.C.
DUANE ALEXANDER, National Institute of Child Health and Human Development, Bethesda, Maryland
MARK BLAXILL, SafeMinds, Tyrone, Georgia
LAURA BONO, National Autism Association, Nixa, Missouri
SOPHIA COLAMARINO, Autism Speaks, New York
ERIC FOMBONNE, McGill University, Montreal, Canada
STEVEN HYMAN, Harvard University, Cambridge, Massachusetts
JUDY ILLES, University of British Columbia, Vancouver, Canada
THOMAS INSEL, National Institute of Mental Health, Bethesda, Maryland
DAVID SCHWARTZ, National Institute of Environmental Health Sciences, Triangle Park, North Carolina
ALISON TEPPER SINGER, Autism Speaks, New York
SUSAN SWEDO, National Institute of Mental Health, Bethesda, Maryland
CHRISTIAN ZIMMERMAN, Neuroscience Associates, Boise, Idaho

IOM Staff
BRUCE ALTEVOGT, Project Director
SARAH HANSON, Senior Program Associate
AFRAH ALI, Senior Project Assistant
LORA TAYLOR, Senior Project Assistant

*The planning committee was solely responsible for organizing the workshop, identifying topics, and choosing speakers. They were not responsible for the publication of the workshop proceedings.

FORUM ON NEUROSCIENCE AND NERVOUS SYSTEM DISORDERS

FRANK YOCCA, AstraZeneca Pharmaceuticals, Wilmington, Delaware
CHRISTIAN ZIMMERMAN, Neuroscience Associates, Boise, Idaho
STEVIN ZORN, Pfizer Global Research and Development, Ann Arbor, Michigan

IOM Staff
BRUCE ALTEVOGT, Project Director
SARAH HANSON, Senior Program Associate
LORA TAYLOR, Senior Project Assistant

IOM Anniversary Fellow
LISA BARCELLOS, University of California, Berkeley

*IOM Boards do not review or approve workshop proceedings. The responsibility for the content of the proceedings rests with the institution.

Independent Report Reviewers

These workshop proceedings have been reviewed in draft form by individuals chosen for their diverse perspectives and technical expertise, in accordance with procedures approved by the National Research Council's Report Review Committee. The purpose of this independent review is to provide candid and critical comments that will assist the institution in making its published workshop proceedings as sound as possible and to ensure that the proceedings meet institutional standards for objectivity, evidence, and responsiveness to the study charge. The review comments and draft manuscript remain confidential to protect the integrity of the deliberative process. We wish to thank the following individuals for their review of these proceedings:

Lisa Croen, Kaiser Permanente Northern California, Oakland, CA
Gary W. Goldstein, Kennedy Krieger Institute and Johns Hopkins University School of Medicine and School of Hygiene and Public Health, Baltimore, MD
Carlos A. Pardo-Villamizar, Johns Hopkins University School of Medicine, Baltimore, MD
Lyn Redwood, National Autism Association, Nixa, MO

Although the reviewers listed above have provided many constructive comments and suggestions, they were not asked to endorse the final draft of the workshop proceedings before their release. The review of these proceedings was overseen by **Dr. Floyd E. Bloom,** The Scripps Research Institute, Professor Emeritus. Appointed by the National Research Council, he was responsible for making certain that an independent examination of these proceedings was carried out in accordance with institutional procedures and that all review comments were carefully considered. Responsibility for the final content of these workshop proceedings rests entirely with the institution.

Preface

Autism spectrum disorders (ASD) constitute a major public health problem, affecting one in every 150 children and their families. Unfortunately, there is little understanding of the causes of ASD, and, despite their broad societal impact, many people believe that the overall research program for autism is incomplete, particularly as it relates to the role of environmental factors. One reason for that may well be that there have been relatively few occasions that have brought together all the key stakeholders—scientists, clinicians, parents of autistic children, patient advocates, and major sponsors of autism-related research—to engage in a full discussion of autism causality and scientific research priorities.

In response to these challenges, the U.S. Secretary of Health and Human Services (HHS) asked that the Institute of Medicine's (IOM) Forum on Neuroscience and Nervous System Disorders (the Forum) host a workshop that would bring together the key public and private stakeholders to discuss potential ways to improve the understanding of the ways that environmental factors may affect ASD. The Forum provided an ideal setting to facilitate this request, since it is designed to provide its members—representatives from government, industry, academia, and patient advocacy organizations—with a venue for openly exchanging information and discussing critical scientific and policy issues related to nervous system functioning.

Thus, on April 18 and 19, 2007, the Forum hosted a workshop, "Autism and the Environment: Challenges and Opportunities for Research" organized by an ad hoc planning committee. This workshop and its development epitomized what is called by many people "public engagement" by and with the scientific community. Members of the broader public were involved in every aspect of the workshop. The planning committee included not only academic leaders and top government scientists, including three institute directors from the National Institutes of Health, but also four members of the autism advocacy community, three of whom are parents of autistic children. Many of the workshop participants and invited speakers were members of the advocacy community. The result was an activity that fully explored from all angles the range of issues surrounding environmental factors and ASD, and resulted in an array of new ideas for research projects and programs. There is no question that this workshop and its product, this volume, were greatly enriched by this broad participation.

As chair of the Forum and the workshop planning committee, I want to acknowledge the hard work and dedication displayed by every member of the planning committee, Forum, and workshop participants. I would also like to thank the leadership of the IOM and HHS for providing the Forum with the opportunity to host this very important event. This workshop was a huge success, both in helping to identify potential scientific opportunities and in demonstrating the utility of moving from a strategy of public education about science toward fuller public engagement, with science where both sides—scientists and members of the public—listened and learned from each other.

Alan Leshner, *Chair*
Workshop Planning Committee
Forum on Neuroscience and
Nervous System Disorders

Foreword

This workshop originated at the suggestion of advocates for patients with autism. In a meeting with the two of us, they broached the idea of engaging with the scientific community to help shape a new research agenda. The Institute of Medicine's Forum on Neuroscience and Nervous System Disorders provided a neutral venue to bring together key stakeholders—scientists, parents of autistic children, other patient advocates, and major sponsors of autism-related research—specifically to identify scientific opportunities to further the understanding of environmental factors that may contribute to autism.

The presentations and discussions at the workshop identified a number of promising directions for research on the possible role of different environmental agents in the etiology of autism. Equally important was the opportunity for dialogue and the exchange of ideas that took place in an atmosphere of mutual respect and learning.

The payoff will be new directions for scientific research that are more fully informed by different perspectives on the reality of autism. From that, everyone stands to gain.

William F. Raub, Ph.D.
Science Advisor to the Secretary
Department of Health and Human Services

Harvey V. Fineberg
President
Institute of Medicine

Contents

*Throughout various speaker presentations, speakers may refer to slides that can be found online at http://www.iom.edu/?id=42481.

Introduction

On April 18 and 19, 2007, the Institute of Medicine's (IOM's) Forum on Neuroscience and Nervous System Disorders (the Forum), in response to a request from the U.S. Secretary of Health and Human Services, hosted a workshop called "Autism and the Environment: Challenges and Opportunities for Research." The goal of the workshop was to provide a venue to bring together scientists, members of the autism community, and the major sponsors of autism-related research to discuss the most promising scientific opportunities (Box I-1). The focus was on improving the understanding of the ways in which environmental factors such as chemicals, infectious agents, or physiological or psychological stress can affect the development of the brain. In addition, discussions addressed the infrastructure needs for pursuing the identified research opportunities—tools, technologies, and partnerships.

Chaired by Alan Leshner, chief executive officer of the American Association for the Advancement of Science and executive publisher of *Science*, the workshop represented a partnership among members of the autism advocacy community, scientists, and policy makers. The autism community was involved in the early discussions that led to the Secretary's request for this workshop and subsequent sponsorship by the Forum and supplemental sponsorship by the National Institute of Child Health and Human Development, National Institute of Environmental Health Sciences, National Institute of Mental Health, National Institute of Neurological Disorders and Stroke, and the Centers for Disease Control and Prevention. Four of the thirteen members of the workshop planning committee—which was solely responsible for organizing the workshop, identifying topics, and choosing speakers—were members of

BOX I-1

Statement of Task

The Forum on Neuroscience and Nervous System Disorders was established by the IOM to provide an opportunity for continuing dialogue and discussion among representatives of all relevant sectors about scientific and policy issues related to neuroscience and nervous system disorders.

In response to a request from the U.S. Secretary of Health and Human Services, the IOM Forum on Neuroscience and Nervous System Disorders, in collaboration with the IOM Roundtable on Environmental Health Sciences, Research, and Medicine, will host a workshop on Autism and the Environment: Challenges and Opportunities for Research. The workshop will feature presentations and discussions on strategies for research focusing on the potential relationship between autism and an array of environmental exposures. An ad hoc planning committee will organize a public workshop that will focus on the following three questions:

- What are the most promising scientific opportunities for improving the understanding of potential environmental factors in autism?
- What scientific tools and technologies are available, what interdisciplinary research approaches are needed, and what further infrastructure investments will be necessary in the short and long term to be able to explore potential relationships between autism and environmental factors?
- What opportunities exist for public-private partnerships in the support and conduct of the research?

the autism community. Furthermore, a number of members of the autism community were speakers, discussants, and workshop attendees, who reminded workshop participants about their sense of urgency in addressing this serious health issue.

The publication of the workshop proceedings provides the Forum with a broader mechanism to inform not only the membership of the Forum, but also other interested parties about what transpired at the workshop. The workshop proceedings should not be confused with a National Academies consensus report. The proceedings do not contain findings or recommendations endorsed by the National Academies or the IOM, the Neuroscience Forum, or the Planning Committee. Opinions and statements included in the proceedings are solely those of the individual persons or participants at the workshop, and are not necessarily adopted, endorsed, or verified as accurate by the National Academies. What follows in Chapter 2 are the proceedings of the meeting. Embedded in this are important lessons for the reader. Proceedings have been edited to eliminate redundancy and grammatical errors. In addition, workshop speakers were provided an opportunity to edit their remarks to ensure clarity and accuracy of statements. Corresponding PowerPoint presentations may be downloaded from the Forum's website (http://www.iom.edu/?id=42481). To assist in the response to the Statement of Task an index of the scientific opportunities that were identified throughout the workshop has been compiled in Appendix A. Subsequent appendixes include a copy of the workshop agenda (Appendix B), a list of the workshop registrants (Appendix C), and biographies of the Forum's membership, workshop planning committee, and workshop speakers (Appendix D).

Proceedings

Day 1
April 18, 2007

WELCOME, INTRODUCTIONS, AND WORKSHOP OBJECTIVES:

Dr. Alan Leshner

Dr. Leshner: Good morning everyone.

I am Alan Leshner. I am the CEO (Chief Executive Officer) of the American Association for the Advancement of Science and Executive Publisher of *Science* magazine, but I am here in my role as chair of the Institute of Medicine's (IOM's) Forum on Neuroscience and Nervous System Disorders.

I am delighted to welcome everyone. This is the workshop Autism and the Environment: Challenges and Opportunities for Research.

The major purpose of this workshop is to work together to try to figure out how we can do a better job to bring the full power of science to bear on a public problem of tremendous magnitude and tremendous import.

The IOM's Forum on Neuroscience and Nervous System Disorders has the purpose of building partnerships and discussions to further understand the brain and the nervous system, to understand disorders and their structure and function, as well as clinical prevention and treatment strategies.

What the forum does is to bring together leaders from the public and private sectors, including federal agencies, the pharmaceutical industry, advocacy organizations, and the academic community to have conversations about these general critical issues.

In addition, we try to serve an educational function, educating the press, the public, and policy makers about neuroscience and nervous system disorders. One of the mechanisms through which we operate is workshops like the one today that provide a venue for discussion about key challenges and opportunities in the field.

The Forum was asked to host this workshop by the U.S. Secretary of Health and Human Services, and William Raub will speak in a moment to help explain its origins, but it came about as a result of a series of discussions among members of the autism community, the Office of the Secretary, and Dr. Harvey Fineberg, president of the Institute of Medicine.

I do want to specifically thank Kelli Ann Davis, Jim Moody, and Mark Blaxill—Mark has also been a member on our planning committee—who were instrumental members of the advocacy community in all of these discussions.

Let me say a few words about the format for today's meeting. You all have a copy of the agenda and I won't read it to you, but let me just reiterate that the workshop objectives are to look to the future, to look for and try to identify the most promising scientific opportunities for improving the understanding of potential environmental factors in autism, to talk about what infrastructure, what tools and technologies are available and what is needed, what kinds of interdisciplinary approaches are needed and other kinds of infrastructure, investments and then to talk about exploring potential partnerships that are needed to support and conduct autism research.

The format that we are using, if you look at the agenda (Appendix B), is to have a series of speakers in each of numerous settings. There actually are way too many speakers for a normal workshop, but we were unable to figure out how to keep this in proportion and make sure that we covered this very complex issue fully. So, we are going to be rather ruthless in maintaining the organization of the workshop.

Each speaker has been given 15 minutes. We will have one or two minutes for what I will call critical questions of clarification right after each talk, but really we mean critical clarification, not a discussion and not a discourse. Then at the end of each session, we have allocated actually a substantial amount of time for discussion among the session participants and then we have allocated time at the end of the day for continued discussion, but also an opportunity for members of the audience to participate at that time as well.

I would ask those people who are planning to ask questions or develop questions over the course of the day's events or the 2 days' events, please, no statements, no long harangues. This is about questions. This is a scientific meeting and we are looking for scientific opportunities and ways that we can move the science forward. So, I remind you of that, but I will surely remind you of that again.

Again, the purpose of the workshop is to stimulate discussion about how best to move research on autism and the environment forward and I need to make some, I apologize, bureaucratic announcements on behalf of the IOM. One, this is not a consensus conference. We are not, in fact, expecting to come to a consensus by the end of the meeting.

What we are expecting to do is to hear an array of opportunities, an array of needs, an array of challenges that will be identified. A proceeding of the workshop will be written. (NOTE: This is the published document to which Dr. Leshner refers in his remarks.) It will not, again, be an official consensus statement or consensus report of the Institute of Medicine. It is outside our authority as a forum to produce those kinds of reports, but, hopefully, what it will do is set the stage for a research agenda moving forward, and that is really what our goal is today.

I do want to thank the planning committee, which has been so instrumental in this workshop. It could not have been done without the joint activity of people from many different sectors with interest in this.

One request of the speakers before we move on and that is—and I apologize for doing this late, but we are a little bit concerned that with all the discussion and all the talk, it may be a bit difficult to capture each speaker's view of what a major opportunity or a major gap to be filled might be. Therefore, if it is not too late to do that organizationally in your head, if at the end of your talk you could articulate at least one, just so the recorders can write down, here is one potential gap in scientific knowledge or potential scientific opportunity that needs to be filled.

If we can focus in that way toward the future, I think we can make a larger contribution than we could otherwise.

I don't want to take too much time. We are already a bit constrained and I, again, want to thank the members of the planning committee, who did such a wonderful job of pulling this together. I want to thank all the speakers, who will be with us today and tomorrow, and all of you who are participating in this.

Let me turn now to Dr. William Raub, science advisor to the Secretary of the Department of Health and Human Services (DHHS), who just will make some additional comments about the workshop and its origins and its particular charge.

CHARGE TO WORKSHOP PARTICIPANTS

Dr. William Raub

Dr. Raub: Good morning everyone. I add my welcome to Dr. Leshner's. I am delighted at the wonderful turnout.

The journey that brought us here began almost 2 years ago with a protest action directed against the Executive Branch. Parents of autistic children and other advocates were mindful of a reelection campaign promise to eliminate mercury-based compounds from vaccines. Staff of the Domestic Policy Council asked me to host a meeting whereby representatives of the advocates could state their concerns in person. I did so and had the privilege of meeting a group of impressive individuals, several of whom are here today.

I would like to be able to say that the protest event, the follow-up meeting at Health and Human Services, and the subsequent communication from the Council back to the advocates left everyone satisfied. But that was not the case, and deep divisions remain over the matter of vaccine safety.

Nevertheless, several of the advocates asked if I would be amenable to hosting further meetings, whereby they could lay out additional concerns about how the institutions of science have approached the problem of autism. I quickly agreed, having been moved deeply by the quality of the advocates' preparedness, the sincerity of their representations, and the power of their testimony regarding the crushing burden that autism places on not only the affected children, but also the entire family.

That led to a series of meetings with various combinations of representatives from the autism advocacy community—but always focused on what science has or has not done and what more it can or should do. Last October, at one of those sessions, Dr. Harvey Fineberg, president of the Institute of Medicine, joined Mark Blaxill, Kelli Ann Davis, Jim Moody, and me to discuss the IOM's studies of vaccine safety and related activities. Out of that meeting arose the notion that some sort of IOM-anchored, autism-oriented event could make a uniquely important contribution to shaping the research agenda against this dreaded disease. During the weeks that followed, Dr. Fineberg and I posed this generic concept to Dr. Thomas Insel, director of the National Institute of Mental Health (NIMH) at the National Institutes of Health (NIH) and chairman of the HHS Interagency Autism Coordinating Committee. Then we widened the circle to include two other NIH leaders: Dr. David Schwartz, director of the National Institute of Environmental Health Sciences (NIEHS), and Dr. Duane Alexander, director of the National Institute of Child Health and Human Development. This team quickly determined that an IOM-hosted workshop focused on potential environmental factors contributing to the etiology or pathogenesis of autism would make for a highly desirable and value-added contribution to ongoing NIH-based efforts to develop a strategic plan for autism research.

Without equivocation, Drs. Fineberg and Leshner affirmed that the IOM neuroscience forum would host such a workshop. Drs. Bruce Altevogt and Andrew Pope and their staff recruited and facilitated the deliberations of a first-class planning committee. The three institutes that I have already mentioned, plus the National Institute of Neurological Disorders and Stroke and the Centers for Disease Control and Prevention (CDC), agreed to provide the requisite funding.

As indicated by the agenda and the advanced materials on the IOM website, the planning committee tried to ensure that no potentially important environmental contributor to autism has been overlooked or excluded. Although the workshop is not intended to reprise the analysis of the epidemiological evidence related to vaccine safety, the planning committee recognized that vaccine constituents, especially organic chemicals used as preservatives or adjuvants, obviously qualify as environmental agents that warrant attention. In other words, our research agenda should include studies of any and all environmental agents that plausibly might contribute to causing or exacerbating autism, irrespective of the medium of exposure. I am hopeful that the next 2 days will prove to be an important milestone for autism research—not only because this workshop is addressing vitally important questions about the cause or causes of the disease, but also because the agenda is the product of collaboration between advocates for autistic children and their families and the scientific community.

To be sure, other aspects of the autism challenge deserve similar attention, especially the paucity of effective treatments, and autism advocates and the scientific community have much further to go to achieve the full measure of mutual understanding and trust. But our challenge here and now is clear: to step together onto the path to a better day, to set the stage for other important steps to come, and to make other advocates and scientists want to be part of that advancing throng.

Thank you for being here.

Dr. Leshner: Thank you, Bill, and thank you very much for your important efforts in getting this meeting organized, getting it stimulated, and setting the appropriate stage for it. I do want to, again, thank the planning committee. I neglected to mention that the importance of this meeting from a scientific and a public health perspective is reflected by the very large number of members of our forum, who came today, in spite of it not being an official regular meeting of the forum. I really very much appreciate the help and support. Almost the entire forum has come today from many different sectors. I think that is an important statement.

I also want to reiterate William Raub's thanks to Bruce Altevogt, Sarah Hanson, and their colleagues from the IOM, who have done a phenomenal amount of work putting this all together and making sure that it happens.

Let me not take more time, but rather turn to Laura Bono, who has been a member of our workshop planning committee, is a board member of the National Autism Association, and has agreed to bring to the group the perspectives of the advocacy community.

PERSPECTIVES OF THE ADVOCACY COMMUNITY[1]

Ms. Laura Bono

Ms. Bono: I am Laura Bono, founding board member and past chair of the National Autism Association. I have been asked to talk about the perspectives of the advocacy community. My time is short, so I will get right to the point of what many in the advocacy community want and think.

Declare autism a national health emergency under the Public Health

[1]Throughout Ms. Bono's presentation, she may refer to slides that can be found online at http://www.iom.edu/?id=42455.

Act and treat it with urgency. Thirty-six thousand children, who should be living normal lives, will succumb to the diagnosis this year alone, affecting the trajectory of their lives and that of their parents forever. Autism is estimated to cost $3.2 million per child over a lifetime. Using the conservative estimate in the United States of 500,000 children means this epidemic will cost society close to $2 trillion.

Autism is both economically and emotionally devastating to the children and their families. Many families are on the brink of bankruptcy as they struggle to get insurance and the medical attention their children need. Murder/suicides of parents and their autistic children are on the rise.

I can't discuss the perspectives of the advocacy community without citing the failings of the CDC. We believe the CDC has a performance and credibility problem. Their failure to declare an epidemic beginning with the 1989 birth cohort to study the time trend data or to examine the toxic and viral body burdens of children are why we are here today, over 15 years too late.

Julie Gerberding, director of the CDC, said recently in a February 8, 2007, CDC press release when they announced the 1-in-150 rate that she wasn't sure if the rates are truly rising or if they are getting better at studies. "Our estimates are becoming better and more consistent, though we can't tell yet if there is a true increase in ASDs [autism spectrum disorders] or if the changes are the result of our better studies." This denial thwarts research into environmental factors and just isn't acceptable. How many autistic individuals did you know when you were under the age of 21?

Since it is impossible to have a genetic epidemic, literally hundreds of millions of taxpayer dollars could have been appropriately directed to gene–environment and other susceptibility initiatives. Even more could have been spent on learning about the critical mechanisms involved in response to environmental neurotoxicants. We could have been focusing on what changed in the environment and when. We could have been investigating the environmental trigger for years and successfully helping suffering children. We urgently need to begin these initiatives now.

Many in the advocacy community are thankful because starting today, the government is finally going to make environmental research a priority, which will lead to better treatments and recovery. Because if autism is environmental, then it is treatable and preventable. It is no longer hopeless, or lifelong. It is hopeful, with a possible cure.

Recent clinical investigations have identified numerous comorbid disease states in children with autism. These include immune system abnormalities; inflammatory bowel disease; oxidative stress; disordered urine and serum chemistries, including elevated porphyrins; methylation disturbances; increased body burdens of metals, including mercury and lead; chronic viral, fungal, and bacterial infections; and microglial activation in the brain.

Studies must be initiated as soon as possible to increase the focus on the identification of these comorbid disease states. Parents and clinicians alike are reporting that when these problems are acknowledged and treated, it can result in marked improvement in children's learning and behavior. Some children recover completely. This should be a wakeup call to us all.

The research paradigm needs to shift from autistic children are genetically defective to autistic children are sick and treatable. We should only grant money to genetic vulnerability and epidemiology studies that have a clear environmental hypothesis. Research detoxification treatments; identify and validate biomarkers; study biomedical imbalances and treatments that are working; investigate the role of vaccines, including thimerosal, aluminum, and live viruses; research the role of the immune, gastrointestinal, and endocrine systems; and study the recovered children's pre- and post-diagnosis medical files for clues.

Because it is the environment, we need to leave no stone unturned. There is a growing body of evidence implicating vaccine overload, mercury and aluminum from vaccines. Thousands of parents agree with this research. They watch their children regress after being vaccinated. Their autistic children have been diagnosed with heavy metal poisoning and immune system dysfunction and when treated, get better. Regardless of controversy surrounding any theory, we must research and produce successful antioxidant, methylation, and blood-brain barrier chelation treatments, as well as immune system, detoxification, and inflammation interventions.

I want to remind you that you are tasked with setting in motion the crucial environmental research that hundreds of thousands of children are silently waiting for now. The guiding principles should be to pursue research and treatments that will impact the most lives as quickly as possible and follow clues provided by treatments currently working in children. Such an agenda would best be served by a translational research protocol where clinicians who care for children with autism advise research into the most promising areas of intervention.

It is imperative that the working group proceed with urgency and follow the truth wherever it leads. Recovery happens every day to some, but our goal should be for all. We need to accelerate environmental research; demand even more money to address the problem; issue RFAs [request for applications] and have research proposals scored according to autism matrix goals; get answers; interpret them expediently; and continue to work the problem until we beat it.

We can do this. And our hope starts with you. Thank you.

Dr. Leshner: Thank you very much for your very powerful statement setting the stage. I would like to respond that we share the sense of urgency. We share the sense that science and research will be the hope, and we of course share the goal of bringing the full power of science to bear on this public health problem of great urgency in tremendous proportion.

I hope that we will live up to the charge that you have just given us to look at the full array of environmental factors and the ways in which they can cause this disorder, affect its progression and then, of course, the variety of ways in which we can approach it. I think that your point about the need to both inform clinical practice, but also to listen to clinical practice is a very important charge and I assure you that we are planning to take advantage of that and listen to that carefully.

So, I really tremendously appreciate the statement you have just made and I promise you that we will do our best to respond to it.

Session I
Autism—The Clinical Problem: "What Do We Know? What Do We Need?"

Dr. Leshner: I would like to turn to a discussion of the clinical problem and introduce Sarah Spence, who will serve as the session chair. I would like to point out that we are on time. So, don't feel any pressure, Dr. Spence, or speakers in this session.

Dr. Spence is staff clinician at NIMH, where she works in the Pediatrics and Developmental Neuropsychiatry Branch. Thank you.

Dr. Spence: Thank you, Dr. Leshner.

I think we may have one of the most difficult sessions to do, which is to introduce the clinical problem and do it in an hour and 20 minutes and no more. So, I am not going to spend a lot of time on the introduction. I think the most important thing to keep in mind during this session is that it is about what we know and what we need. It is about introducing the main issues to set the stage for a productive discussion later on today and getting a diverse audience, kind of onto a level playing field about what the issues are.

So, to start with the clinical problem, I am going to introduce my boss, Dr. Susan Swedo, the chief of the branch that I work in at the NIMH.

CLINICAL OVERVIEW: HOW CAN THE CLINICAL MANIFESTATIONS OF AUTISM SHED LIGHT ON POTENTIAL ENVIRONMENTAL ETIOLOGIES?[2]

Dr. Susan Swedo

Dr. Swedo: Since I only have 15 minutes today to describe all of autism to you and why we believe that the environment plays such a crucial role in this disorder, I'm going to be using videos to show you in a few seconds what it would take me a very long time to try to explain.

---

[2]Throughout Dr. Swedo's presentation, she may refer to slides that can be found online at http://www.iom.edu/?id=42456.

Autism is characterized by two areas of deficit, deficits in social interactions and communication deficits. It is also defined by an excess of repetitive behaviors or fixated interests. Now, these fixated interests and repetitive behaviors are not usually present during the very earliest stages of the illness and increase in time as the child becomes older.

As we know, autism is a developmental disorder. By definition, symptoms must appear before age 3 years and affect development. The crucial thing is that the development affects the symptom expression and the symptom expression also affects development. Since this is a disorder of social communication, which is essential to all development-al interactions, autism can quickly take you very far off of your expected trajectory. Autism is one of several pervasive developmental disorders (PDDs), which are now commonly called the autism spectrum disorders.

I think that expanding the continuum to include all pervasive developmental disorders as "autism" is a bit confusing and dilutes the meaning of the term, so I am asking that we keep our focus today on those children who meet full criteria for autism.

Rett disorder is caused by a genetic mutation, which leads to symptoms very similar to autism and, in fact, until we knew what the gene was, girls with Rett disorder were included in the autism group. Since the gene has been identified for Rett disorder, it is now considered to be separate from other autistic disorders. Similarly, childhood disintergrative disorder presents with symptoms of autism, except that the children don't begin to regress and lose their skills until after age 3.

Here is an example of the social deficits in autism. One of the crucial components of social interactions is joint attention—being able to pay attention to things that are of interest to others. (Video shown of a child performing a task of joint attention.) Here you see a normal volunteer from our lab. His reward is a bunny and he is very clearly excited and he tries to share that excitement with the examiner.

Here is a 4-year-old girl with autism performing the same task. The bunny is behind the examiner again. You see the examiner saying "Look! Is it a bunny?" but the child is oblivious, preoccupied with other thoughts. I am going to replay that section of the video and ask you to also watch the repetitive behaviors that she exhibits. Notice that the child pulls her hands into her sides. Then when she gets excited, there is a repetitive motion.

Another common social interaction is shared enjoyment. (Video shown.) The young child with typical development says, "Wow! Look at that!" when shown the bubble gun. He invites his mother to share in his enjoyment of the new toy before asking if he can have a turn operating it. It is easy to see how excited he is by the toy. He uses gestures to make his needs known, as well as his verbal comments.

(Next video is shown.) Here is XXXX,[3] a little boy of about the same age, with fairly severe autistic symptoms, presented with the same bubble task. He clearly sees the bubbles, is interested in them. The examiner gives him every cue she can to get him to ask for more bubbles, but he doesn't. He just seems terribly confused and somewhat upset.

(Video segment.) Here is an example of an autistic child's perseverative behaviors. You probably have heard about the autistic children who spin the wheels of the bus, rather than playing with the bus as it is intended. Here the pop-up toy has become an instant area of fixated interest for him. He isn't playing with it as intended, but rather, chooses to repetitively open and shut one of the doors. The examiner is trying to get him to look over at the bunny. But he is not willing to attend to anything but the pop-up toy. Even when she gently takes the toy away, he remains fixated on the spot where it was sitting. So, this child demonstrates both deficits in social communications and an excess of repetitive behaviors.

The causes of autism that are known are mainly genetic. About 10 percent of children diagnosed with autism have been found to have a genetic cause. Less than 1 percent have been attributed to teratogens, such as valproic acid or thalidomide. That leaves about 90 percent of the kids, or 9 out of 10, for whom the cause is idiopathic, meaning we just don't know. That does not mean that there is no known cause. It just means that the cause is not known.

When autism is related to a genetic defect, the pathogenesis is relatively "simple." Even then, there is a great deal that happens between the genetic mutation and the manifestation of neuronal dysfunction and/or damage. But when something in the environment is causing the symptoms, it is even harder to make a direct link. But the working model is that environmental factors, in a genetically susceptible population, lead to neuronal dysfunction and/or damage and the symptoms of autism.

The tricky thing about that pathogenic model is the fact that it has so

---

[3]Out of respect to privacy for the family, the name of this individual has been replaced with "XXXX."

many stages, each of which is actually broken down into many, many more steps. So, for the purposes of the conference, you are going to be hearing a lot about genetic mechanisms that might create vulnerabilities, about the environmental factors that trigger the symptom onset, and even though we'll be addressing individual parts of the diagram, we need to keep the larger picture in mind at all times.

Potential environmental triggers that have been suggested are numerous. They include the toxicants, which will be discussed by Isaac Pessah; the infectious agents, which Ian Lipkin will be speaking to in a later session; and household exposures, such as household chemicals and cleaning products. The household exposures are one of the areas of study for the NIEHS-sponsored CHARGE (Childhood Autism Risks from Genetics and the Environment) study and the CDC-sponsored CADDRE studies (Centers for Autism and Developmental Disabilities Research and Epidemiology).

Food, dietary supplements, and vitamins and minerals may also be involved in autism. If you think about how we eat today, compared to how we ate in the 1950s, it is mind boggling how many changes there have been. Of particular interest have been changes in folic acid supplements, and the utilization of aspartame, because both have been associated with other neurologic conditions.

Additional environmental factors include drugs, medications, and herbal remedies. For example, as a pediatrician, I know that there was a dramatic change in the treatment of children with fever following the Reye's syndrome epidemic. And practice guidelines required a switch from giving children aspirin following vaccinations to prescribing Tylenol and/or ibuprofen. We don't know what the effect of that might have been, but it is certainly an area for investigation.

Other medical interventions that might play an etiologic role include the use of ultrasounds during pregnancy, and the administration of vaccines—not just the contents of those vaccines, but also the increasing number and the immunologic challenges that are faced by our children today, in comparison with previous generations.

Technological advances include the ultrasounds, but also microwave ovens, cell phones, and everything else. So, you really end up with an overwhelming array of environmental factors to consider because in essence, everything encountered by the mom, the dad, and the child could be a potential environmental trigger.

There are some clinical clues that suggest that the environment is playing a role in the etiology of autism: first, the association with the teratogenic agents is a direct cause-and-effect relationship; second, the reported prevalence of autism is increasing at dramatic rates; and third, the fact that the symptoms frequently have their onset between 12 and 18 months of age (not at birth). I think this is the thing that the parents see as the most compelling evidence that there must have been an environmental trigger. They tell us, "My child was healthy and then he wasn't"—something must have happened in between.

The change from typical development to autism certainly may have been the result of an environmental exposure, but we have to keep in mind the fact that many disorders that are genetically based do not present in the first year or even 2 years of life. Sickle cell disease is a prime example. In addition, there are disorders like Rett syndrome in which the girls are developing normally until about 12 to 15 months of age and then have a regression and lose their skills. So, I think that the age at onset of symptoms in autism is an important clue, but it isn't evidence on its own.

Medical comorbidities may also provide information about environmental factors in autism. For example, within the past few years, there has been increasing attention to the link between autism and immune dysfunction that suggests a common environmental exposure is increasing prevalence rates for both autism and autoimmune disorders. We will hear more about that during this workshop as well.

A request has been made that we start paying attention to the response to treatments that are being given to these children in order to find clues to the original etiology of symptoms. Many parents and practitioners are finding that symptoms can be dramatically improved or eliminated by a variety of biological and dietary interventions. At the NIH, we are attempting to do systematic studies of some of the more commonly used treatments, because open-label trials and anecdotal reports of benefit can be very difficult to assess because the child is developing naturally during that same period of time.

The regressive subtype of autism is one of the most clinically compelling pieces of evidence for environmental triggers. The regressive subtype of autism is actually regressive "subtypes," just as there are multiple autisms. For most children with regressive autism, they develop normally until about 12 to 30 months of age, when they begin to lose the language they have acquired and stop interacting socially. However, 12 to 30 months of age is a tremendous span in development, and suggests that even within the regressive group, there is likely to be a significant amount of heterogeneity.

Fifteen to 50 percent of children with autism will have regressive features, depending on how narrowly you define "regression." If you take the strictest definition, which requires that the child has at least 10 words and loses those, then the proportion is closer to 15 percent. To date, the prognosis for the regression group is reported to be particularly poor.

Of note is the fact that regression can be very acute. We have already seen children at the NIMH clinic who were developing normally, became ill, and within a few weeks had lost all of their verbal and social skills. For most children, the process is slower and subtler; it is a painstaking process to find out how they were developing at each developmental phase and to begin to pinpoint the area at which the regression occurred.

The final caveat in consideration of the regressive autism subtype is work from Dr. Geraldine Dawson and her colleagues at the University of Washington which shows that for many of the children, development wasn't completely normal before the regression occurred, but there is still a very obvious loss of acquired skills. Here is an example of a little girl who had a clear regression. She is the one that you saw with the self-stimulatory behaviors and the lack of attention. Here she is at 6 months of age. Her dad calls her name, and she gets a huge smile. Here she is at her 1-year birthday party. Again, her father calls her name, and see if you can tell when he says it. You can't, can you? So, she had already lost attention to her name. By the time she is a year and a half, he is shouting her name repeatedly, and she is completely oblivious to his presence. She had also lost words during this period. As you can see in the videos, the regression is profound. The family describes it as having their daughter "stolen" from them by the autism. I think that is a superb description to keep in mind of the regressive subtype. The child is developing on an expected trajectory and then falls off completely.

Certainly in regressive autism, the hunt for the environmental trigger should take prominence, but how do we trace back from the clinical picture to that environmental trigger? As I said earlier, it is complex. Each of these cartoon boxes has multiple stages, multiple phases, and multiple levels to be investigated—it is a huge task, but it isn't hopeless.

I was asked to tell you what I think we need to do to find these environmental factors. First, we need a standardized definition of autism and related disorders. We really need to be dealing with as clinically homogeneous a group as possible, because within that homogeneous group, we are going to find biological heterogeneity. We already know this from all of the other medical disorders of childhood, and particularly from Type 1 diabetes and leukemia, where knowing exactly what the clinical picture looked like helped us to get to the pathophysiology.

We need brain pathology. As we had our planning conference calls, it became very clear that until we know what is happening in the brain, there is not much point in trying to figure out when or where the trigger occurred.

It would be helpful to have incidence data from populations with disparate risk factors. If we could look at developing nations and their rates of autism, we might be able to find clues to environmental triggers here in the United States and elsewhere in the industrial world. In order for such studies to be meaningful, however, we need to use the same diagnostic criteria for each time and place. It is very clear from work being done by international epidemiologists that if you change the diagnostic cut-off scores by just one point, the prevalence rates change dramatically. Obviously, the same thing would be true for the incidence data and would complicate any international comparisons.

We need systematic evaluation of anecdotal case reports as we already know from genetic disorders that it is the exception that ends up proving the rule. So, we need to start looking for those exceptions and studying them in depth. At the same time, we need to be doing randomized control led trials of novel therapeutics, using reliable, valid, developmentally appropriate and change-sensitive outcome measures—such measures still need to be developed. And finally, we need identification of clinically meaningful subtypes, perhaps by identifying unique ages of onset, similarities of clinical presentations or associated symptoms, or by identifying a group with similar developmental or clinical trajectories.

Since I am out of time, I will stop and take questions.

Dr. Spence: So, the next 2 or 3 minutes we can use for questions directly related to Dr. Swedo's talk or else we can move on.

Dr. Swedo: Since there don't appear to be any questions, I am going to spend the next 3 minutes talking about PANDAS (Pediatric Autoimmune Neuropsychiatric Disorders Associated with Streptococcal infections) and how we at the NIMH were able to use clinically meaningful subtypes of obsessive-compulsive disorder (OCD) to go from the unique clinical presentation to the environmental trigger, and meaningful treatment and prevention strategies in a relatively short period of time. Our hope is that we will be able to find a similarly informative subgroup of children with autism.

The PANDAS subgroup differs from other children with OCD in that it has a very abrupt onset and an episodic course, in which there are periods of both relapse and remission. Boys predominate in this young population of children. When the children are acutely ill, they have developmental regression, social isolation and aggressiveness, emotional lability, sensory defensiveness, sleep difficulties, and choreiform movements. The symptoms are found in many children with autism, as well as in Sydenham's chorea, which is the neurologic manifestation of rheumatic fever. The association between obsessive-compulsive disorder and Sydenham's chorea is what led us to suspect that strep bacteria might be the environmental trigger for the abrupt-onset form of obsessive-compulsive disorder. A decade of research suggested that the presence of untreated strep bacteria in a genetically susceptible host could cause an abnormal immune response and lead to clinical manifestations of obsessive-compulsive disorders and tics. We already knew that only a few of the 120 strains of strep were capable of producing rheumatic fever, and that not all children were susceptible to the poststrep complications. In fact, only about 1 in 20 families was susceptible to rheumatic fever. It seemed like a difficult model to investigate—not all strep infections could cause symptoms and not all children would be affected, so there would be many false starts and dead ends.

However, by starting with this model, we were able to borrow from the experience with rheumatic fever eradication, and conducted a controlled trial of antibiotic prophylaxis that showed beautifully that preventing strep infections was capable of preventing neuropsychiatric symptom exacerbations. By giving antibiotics to prevent strep, we were preventing episodes of OCD and tics. The slide shows the results of the trial for the first 10 patients—on the left side of the red line is the year prior to study entry and on the right side is the year of antibiotic administration; just visually scanning the data, you can see that there are fewer symptomatic months (represented by the bars) during the year of antibiotics administration. The summary data showed that the children went from having two strep infections on average per year to zero strep infections, and that they went from having 2.4 to 0.7 neuropsychiatric exacerbations during that same period. What isn't shown here are the follow-up data demonstrating that continued antibiotic prophylaxis has rendered over 75 percent of these children asymptomatic.

The genetically susceptible host allows us to develop trait markers and susceptibility markers. We had great hope for a short period of time that the D8/17 marker would serve as a susceptibility marker for the PANDAS subgroup. Unfortunately, the original monoclonal antibody clone was lost and we haven't found one that has equal sensitivity and specificity, but the hunt goes on.

The postulated abnormal immune response led to two lines of investigation. First, the search for a disease marker which would reliably distinguish children in the PANDAS subgroup from others with OCD, and the development of immunomodulatory treatments for severely affected children.

Dr. Madeline Cunningham and Christine Kirvan have been the heroes in the search for disease markers. They have demonstrated that cross-reactive antibodies recognizing the strep cell walls also recognize neurons within the basal ganglia and that the titers in the Sydenham's chorea group (shown on the left side of the graph in the red squares) are much higher than those in the PANDAS subgroup, but the PANDAS children are significantly higher than the normal controls, and most importantly, acute and convalescent titers are dramatically different in both Sydenham's chorea and PANDAS. Thus, the antibody titers may be useful not only in identifying PANDAS versus non-PANDAS cases, but also in following disease progression and response to treatment.

We also conducted a placebo-controlled trial of immunomodulatory treatments in which plasma exchange and IVIG (intravenous immunoglobulin) therapy both were effective in reducing symptom severity by more than 50 percent in the first month following treatment, whereas placebo had no discernible effects. In conjunction with relieving symptoms, the immunomodulatory therapies also reduced the size of the abnormally enlarged caudate nucleus, as seen in this individual.

So, for PANDAS, we were able to identify a medical model for disease etiology and to use that model to prevent symptom onset by preventing strep infections. We were also able to identify the genetically susceptible host and develop markers of disease activity, and even develop treatments that were effective in eradicating the neuropsychiatric symptoms. I would challenge us to try and do the same thing in autism. It is not going to be easy, but if we start with clinically meaningful subtypes of autism, we will be able to identify the etiologic triggers and keep them from doing harm. We will also be able to identify biomarkers of genetic susceptibility and develop diagnostic tests that will identify vulnerable populations. And, of course, our ultimate goal is to move from the clinical observations to developing new methods for prevention and cure.

Thank you.

Dr. Spence: Thank you, Dr. Swedo.

Next we have Dr. Pat Levitt, who is a professor of pharmacology at Vanderbilt.

GENES AND THE ENVIRONMENT: HOW MAY GENETICS BE USED TO INFORM RESEARCH SEARCHING FOR POTENTIAL ENVIRONMENTAL TRIGGERS?[4]

Dr. Patrick Levitt

Dr. Levitt: I am going to provide for you a neurobiologist perspective on where we are in terms of genetics and what some of the opportunities are in terms of genetics and designing the kinds of research we might be doing to understand gene–environment interactions. The first slide basically depicts the fact that we all understand—complex genetic disorders are complex.

Complex genetic disorders are complex and what we are trying to understand are the combination of risk alleles, variations in gene sequences or in copy number of specific genes which, in combination, end up underlying risk or, in fact, directly perturb brain development that ends up generating the three core symptoms that are diagnostic of autism spectrum disorders.

You can see in the diagram that for any disorder, a combination of risk alleles may be correct, but there may be an intermediate phenotype rather than the features of the full disorder. We know that of the three major core symptoms that are used for an autism diagnosis, dysfunction in any one of these domains can run in families. There have been large twin studies to look at heritability independent of the autism diagnosis itself.

The diagram also shows that the correct combination of risk alleles

[4]Throughout Dr. Levitt's presentation, he may refer to slides that can be found online at http://www.iom.edu/?id=42457.

might require specific environmental factors in order for the full-blown disorder to be expressed. There also are issues of incomplete penetrance where you may have the correct combination of genetic risk, but for reasons unknown, an individual has modifier factors that reduce the impact of the risk alleles. This means that one does not express the disorder.

So, I want you to keep something in mind. I take this from Daniel Weinberger, who studies schizophrenia at the NIMH and he makes this point, I think, very well. Genes are involved in the assembly of specialized cells to perform specific functions. Thus, there are no "social behavior" or "communication" genes. If we are looking for those, you might as well stop now because they don't exist. Genes don't know about social behavior. They don't know about communication. What they know about are assembling tissues and cells to perform specific functions, and when there are mutations or changes in the sequence of those genes that affect function or expression levels, or differences in the copy number of those genes, we see alterations in the assembly of cells and the specific functions that they underlie.

So, what do we know from a genetics perspective? Well, there have been three approaches used: (1) linkage studies that look for excess sharing of genomic regions on chromosomes that track with the disease; (2) allelic association studies, where we look for excess sharing of alleles; this is accomplished by studying single nucleotide polymerphisms (SNPs), in which a single nucleotide is changed, or differences in microsatellite sequences at a single locus; and (3) a defined copy number variation (CNV), where we look for submicroscopic changes (thousands of bases, rather than macroscopically identified millions). CNVs thus are not obvious changes, such as chromosomal rearrangements, but submicroscopic changes that alter chromosome structure, which could be either deletions or duplications. Keep in mind that most of the chromosome is not occupied by sequences of bases that encode the transcript that will be translated into protein, but rather encode regions whose functions we really don't understand, but we think may be involved in regulation of gene expression.

I want to mention here some of the previous and current caveats to what we know in terms of ASD and genetics. You need to keep these in mind as you read genetic studies to determine the degree to which you can rely on the findings and conclusions. First, there may and are likely ascertainment biases. This means that the subject population that has been studied genetically may not necessarily be broadly representative, or perhaps they are broadly represented, but they represent one small domain of the spectrum. Second, until recently the sample sizes used in studies typically were small and underpowered. Why is this a problem? Well, as my friend, Ted Slotkin, tells me, if you do enough comparisons, you will find something. For genetics, this means that if you try to find an association between many different SNPs and a disorder, eventually you will identify some relationship statistically—but one needs to correct for what we call Type 1 error, that is, false-positive results. There are debates regarding the best ways to correct, and many earlier studies may not have corrected at all, leaving us with nonreplicable findings.

Third, the accuracy and completeness of the diagnosis and characterization of the phenotypes are essential to understand who you are studying in terms of a cohort to be used in a genetic study. If this is not done at a high standard, an already heterogeneous disorder like autism becomes even more difficult to study genetically because the study population may be diluted with poorly defined subjects. Fourth, in the past there have been issues with technical quality control; that is, the quality of the assays used to identify SNPs and other changes. This is becoming a nonissue as technology advances. Finally, and perhaps most important, there is for the most part a lack of assigned gene/variant function, in which the polymorphism does something to gene expression or gene function. Keep in mind that this concern comes from a biologist. One may identify a variant associated with autism, but if it is not a coding variation that would clearly change the coding of an amino acid, what does it do? You are stuck with that finding in terms of translating that to a biological substrate for the brain changes that may underlie the disorder.

So, what do we know about linkage? Well, the most recent autism genome project consortium identified a modest signal on chromosome 11p, and this is being followed up. In addition, by doing some data filtering, a few other loci seen in previous studies were seen, including regions on 2q and 7q. From previous work, there are in the literature dozens of other reports of linkage, but the bottom line is that with disease heterogeneity, as we have seen in schizophrenia, for example, and other disorders, the linkage signals are generally relatively small and there may be difficulty in replication from study to study. This is telling us something about the disorder, that there is locus heterogeneity. Thus, there are likely to be many different genes or combinations of risk alleles that may underlie ASD.

Regarding CNV, there has been a lot of discussion over the last month about two studies, the AGP (Autism Genome Project) study and a study out of the Cold Spring Harbor group, essentially identifying somewhere between 8 and 10 percent of the individuals in their study having CNVs. The findings are exciting, as CNVs have been implicated in other disorders (e.g., certain cancers), but the findings are not without issues. There is no overlap in terms of chromosomal sites, as far as I could tell, between what was found in the Cold Spring Harbor study and in the AGP study and one does not know the biological significance yet. We are going to talk about that later perhaps, I think, in terms of what CNV might contribute to this disorder.

Regarding rare mutations, we know that there are loss of function mutations that have been identified in a single individual that was part of a genetic study, or even in individual families in which there is an autism diagnosis. I have listed some of those genes up there. The reason I list those is because it turns out that a number of those mutations are found in genes that at least biologically have some things in common; they are involved in synapse formation and function. Keep in mind, however, that rare mutations generally do not translate into genetic variations across large segments of the affected population. They are important in trying to understand the genetic contribution to the neurobiological disruptions.

Genetic syndromes with co-occurring ASD diagnosis have often been overlooked in the past. My friend Art Beaudet talks about these all the time. Disorders such as Fragile X, Rett, Angelman, and Timothy have a relatively high prevalence of co-occurring autism diagnosis. In addition, there are some common themes in terms of the neurobiological changes known to occur in each of these disorders, related to the changes in neural development. Keep in mind that genetically the causes are quite distinct from each other, but the high co-occurrence suggests that there may be many genetic routes to impact negatively the three core functions used to diagnose ASD.

The literature is also replete with reports of association of common risk alleles with ASD; that is, gene variants have been identified from standard association studies that give us some clues regarding the impact of common variants on genetic risk. I have listed some of those on this slide: (1) nonfunctional risk alleles, meaning that there has been a change in the sequence of the gene, but we don't know what that sequence change means. I have listed some of those genes up there. The neurexins, the GABA beta 3 subunits and Gral-2 and some risk alleles that have been identified, but not necessarily replicated in every study; and (2) functional risk alleles, that is, variants that have been identified that either change the function of the gene product or change how much of the gene product is actually produced. The promoter region of the reelin gene is one example, and I have placed a red circle around Met, a finding from my laboratory that I will tell you about in a moment.

I have posed some questions related to the influence of genetics on autism expression that might also be retitled "Gaps in Knowledge" (1) How much of genetic risk is due to direct impact of mutations on brain development? (2) How much of what we are talking about in terms of genetic influence is actually the combination of genetic mutations changing the trajectory or course of neural development in wiring the brain up? (3) How much of the risk is due to direct impact of mutations on peripheral functions, that is, other organ systems that influence brain development? I raise this as a possibility because we know that peripheral organ development and brain development are linked physiologically. (4) How much of the risk is due to genetically established sensitivities to environmental perturbations? We know this exists experimentally, but we really don't understand it in the clinical population. (5) How much of the phenotypic heterogeneity of individuals with ASD are influenced combinatorially through genetic and environmental factors? That can be viewed as my red herring question.

So, here is my concept of where we are with understanding autism brain pathophysiology. I am being facetious, but that is a thimble in case you didn't recognize that blurry image. In essence, we know very little about the changes in brain development and brain organization that underlie ASD. That is a real problem in trying to understand the causes. Genes, environment, or both? How can you answer any of the questions I posed without knowing what exactly is disrupted in terms of brain architecture and development? Part of the problem, in my opinion, has been that the gene–environment debate has been held in isolated silos, that is, separated disciplines in which there is rare exchange of ideas. The silos, or disciplines, need to interconnect. This harkens back, and we talked about it on the conference call among presenters, to when developmental neurobiologists spent an enormous amount of time trashing each others' work because one was either in the "nature school" or the "nurture school" regarding brain development. Of course, that was silly because we know that the brain is built through a genetic blueprint that takes information from the outside world and utilizes it to direct the developmental course to wire up circuits. This gene–environment interaction is one of the unique properties of the brain. So, of course, regarding ASD, it is not genetic versus environmental, irrespective of whether you think there is a principal cause that is genetic or environmental. Because ASDs have at their core disrupted brain development, in terms of etiology, both genetic and environmental influences must play roles because this is in the basis for brain development.

So, here is a concept regarding what we might do to address mechanisms: Translational approaches that incorporate multiple technical strategies. There are a number of different strategies in which we are trying to link these domains experimentally. One approach is to focus on neurodevelopmental genes that have been characterized for altering the assembly of circuits that are likely to be disrupted in individuals who develop autism; it does allow investigators to move freely between animal models in which the biological functions of the genes are studied, and going back and working with human geneticists to try to determine whether there are meaningful relationships that would make sense in terms of variations of that gene that might underlie partial risk for ASD.

One also can begin from human genetic research data and develop model systems that probe biological functions, trying to make sense in terms of what has been identified as a variant associated with the disorder that carries genetic risk. I would suggest that it doesn't necessarily help to knock out a gene in a mouse if the variant that has been identified in the human genetic research is not a complete null, but rather a variant that alters protein function or levels of gene expression. Genetic knockout studies may generate some very interesting biological findings, but these may not necessarily be relevant to the pathophysiology of ASD. Of course, with model systems, such as genetically engineered mice, you can do experiments. You can manipulate the system both genetically and environmentally at different developmental ages. I have diagrammed an example, in which one can expose genetically manipulated mice to different environmental factors that we know change the course of trajectory and development. The impact of exposure may be influenced by genetic variation and you can design experiments to do this in developing model systems.

So, I just want to highlight for the last minute or two what we have done in our laboratory. There is only one data slide, and it summarizes work published in 2006 in the *Proceedings of the National Academy of Sciences*.

We took the approach of studying the role of a gene in brain development and then extending these to human genetic studies for several reasons. We were examining the role of a tyrosine receptor, Met, in cerebral cortical development. Met actually has been the focus of thousands of scientific studies because its dysregulation is implicated in certain kinds of cancers. It turns out that this gene is expressed in the brain during development and is important for a number of different processes, including cell migration, development of excitatory and inhibitory neurons, synapse formation, and myelination. We were studying Met in an animal model. The brain architecture changes we found when we manipulated levels of Met expression, together with long-term changes behaviorally, paralleled changes in ASD. We also realized the Met is located under a linkage peak on chromosome 7 in humans, a region implicated multiple times in studies of ASD.

The major finding is that we identified an SNP in the 5' region of the gene that controls how much of the gene is expressed. We showed experimentally that it reduced how much of the Met gene is expressed, and we believe the mechanism for this is due to the 5' SNP associated with ASD reducing the ability of two transcription factors to bind to this region of the gene. Transcription factors are proteins that control how much of a gene is turned on in specific locations and at any particular time during development.

Thus, the Met variant that is strongly associated with ASD actually had a functional outcome. It changed how much of the gene was actually produced. Met is involved in brain development, but we also thought more broadly about this when we were debating about doing the human genetics studies. Met is also involved in gastrointestinal repair, in immune response regulation, and some other peripheral functions that are consistent with the co-occurring medical issues that are described clinically for individuals with ASD.

We spent a lot of time with clinicians to talk to them about whether this made sense because it is not a small number of children who have co-occurring medical conditions. Though still unsettled, it may be a relatively large number. These detailed delineations of the population are telling us about disorder etiology and perhaps even the biology as well.

It turns out that the transcription factor impacted by the ASD-associated variant in Met is SP1, which happens to be a transcription factor whose binding to DNA is disrupted by a number of environmental toxins. So, here one can see that the possibility of combining environmental toxin work with this variant in a humanized mouse model, for example, or introducing the humanized mutations in cells, opens up the possibility of studies that examine combined genetic and environmental influences.

So, for one example, we have actually shown that if you expose cells that have either the G or C (ASD-associated) variants of this gene to BaP, which is a common environmental toxin, levels of gene expression are reduced quite dramatically for both the common and ASD-associated variants. Keep in mind that the common SNP (G) results in more than double the amount of gene transcription in the cells than the ASD-associated variant. I've added here a hypothetical threshold for when a disorder is expressed. If the toxin reduces levels of Met expression for both the G and C variants, but the C variant starts out lower, the environmental exposure will result in even lower levels of expression that reduce below the threshold. In this example, even with BaP exposure, expression of the gene with the G allele still does not drop below disorder threshold. Thus, BaP does not directly cause the disorder, but has differential effects due to genetic variation.

So, what do we need to do in the future? I've listed some suggestions here. We need to increase subject ascertainment, character-izing populations in great detail, which will allow geneticists, psychologists, and neuroscientists to stratify groups more accurately to determine if certain phenotypes are associated with specific genetic variants, including SNPs, CNVs, and other genetic changes. Given that we all agree that we need to be very careful about how we phenotype in doing the genetic studies, it simply doesn't make sense to start out with a cohort of 1,000, because by the time you stratify based in different characteristics of ASD, or even life history, the study will be underpowered.

Deep sequencing to identify more functional variants will be important to pursue. If we are going to translate the genetics to more than just associations or statistical arguments, we have to translate the findings to biologically relevant changes. Thus, functional characterization of the variants is a very high priority.

There needs to be continued wise investments in model systems that will allow us to pursue gene–environment influences more rigorously than can be done in human populations. Finally, if we are going to understand functional etiology of ASD, if we are trying to identify the genes that underlie risk, and we are searching for environmental factors that cause changes in brain development, we need to know where these candidate genes are expressed in the developing human brain, and where these environmental factors have their impact. There is a difference between mouse and human brains, and it is essential to keep in mind that one cannot always extrapolate findings between species because of fundamental differences, particularly related to brain areas that simply are not represented in the mouse, but which may be at the heart of ASD. For this type of information, there is an enormous gap in terms of understanding where key genes might be playing a role in neurodevelopment, and how their perturbation may impact the core features of ASD.

Thank you.

Dr. Spence: Thank you, Dr. Levitt.

Now, we will move right on to Dr. Isaac Pessah, who is the director of the Children's Center for Environmental Health and Disease Prevention at the University of California–Davis, at the M.I.N.D. Institute. He will give a toxicology talk.

HOW MAY ENVIRONMENTAL FACTORS IMPACT POTENTIAL MECHANISMS IN HUMANS?[5]

Dr. Isaac Pessah

Dr. Pessah: Thank you. Pat Levitt really summed up very nicely the complexity of the heritability of autism and autism spectrum disorder. What the last 25, 30 years have taught us is that any chromosome in the genome has multiple linkage sites, whether they are replicable or not, but the fact is that many, many genes may be involved in conferring susceptibility to autism, and to a toxicologist, one would say if there are that many genes involved and more than one gene in any individual that is susceptible to autism, environment must play a factor.

So, what do we know about the scope of the problem in terms of how

[5]Throughout Dr. Pessah's presentation, he may refer to slides that can be found online at http://www.iom.edu/?id=42458.

environment may influence outcome? Well, here are some data that are in the broadest sense not necessarily hard-wired data and very plastic in how you interpret them, but of about 4 million births per year in the United States, about 120,000 show major birth defects and these include structural defects, growth retardation, functional deficits. We believe that this underestimates the problem because most neurological and behavioral problems are not diagnosed until early childhood or young adulthood. So, this is essentially an underestimate at birth. At present, the cost and the causes of the majority of the developmental defects are not understood, but it is believed that about 3 percent of all developmental defects may be attributed to exposure to toxic chemicals and 25 percent of all developmental defects may be due to a combination of genetic and environmental factors, where the person's genome essentially confers increased susceptibility to the environmental hits that occur both during gestation and postnatally.

Now, I wish I could tell you that the environmental issues, the chemicals, are simpler to address. In fact, they are much more complicated. These are very old data. They were released by a National Toxicology Program Report back in 1992 that was mediated through the National Academies. Essentially, this thorough review of the literature came to some rather startling conclusions—that at the time there were greater than 53,000 commercially important chemicals in use, and approximately 80 percent lacked adequate toxicity testing for risk assessment.

This is especially true in the vein of neurodevelopmental toxicity testing. You would think that pesticides are more highly regulated and, therefore, we have more information on them and, in effect, we do, but still they concluded that 64 percent lack adequate data for risk assessment. Cosmetics are not any better, and food additives—here we are talking about intentional food additives that may have unintentional consequences.

One example that I like to use is the fact that high-fructose corn sweetener is processed through reagents that are generated by chloralkalide plants. Chloralkalide plants use a mercury cell process and now NIST (National Institute of Standards and Technology) and some other labs, including ours at Davis, have identified very low levels of mercury within high-fructose corn sweetener, very low levels. Yet, one doesn't know how to use this information for risk assessment because of volume and we actually don't know the form.

I have updated information that there is now at the Environmental Protection Agency (EPA) a very active program to try to identify which chemicals we should, in fact, be prioritizing for additional toxicity testing so we can improve our risk assessment.

The little pie chart that probably doesn't appear on your monitor essentially shows that there are hundreds, if not thousands, of chemicals in various environmental samples, including foods, water, and so forth, that we really have begun to prioritize in terms of trying to understand risk assessment.

What is really needed is information about the additivity of various combinations of these compounds that may cause actually synergism, where each compound doesn't cause an effect, but together they have a much greater effect. Some of these exposures may antagonize each other, and also very important to risk assessment is the relative timing of exposure.

So, let's get back to autism. Well, I would like to propose, and I think others have as well, that autism is really a multisystem disorder. We have focused on the developing nervous system, but because of the number of genes involved and the heritability pattern and the fact that greater than 90 percent are idiopathic, we have no clue how it is caused; we should assume that children who are susceptible to autism may actually be more adversely impacted by environmental exposures than the typical child.

What are the possible mechanisms involved? This was a large task for me. I am going to generalize and then give you two examples. One is a simple example that will provide a framework for additional studies and one is more complicated. If we look at this kind of hypothetical curve here, where we are looking at the percentage of kids having adverse effects that we can actually measure as a function of a particular cadre of toxicant exposure, this could be a single chemical or a complex mixture, even highly inbred rats, even cell lines will show you variability about the meaning for the adverse outcome that you are measuring.

So, you will have hypersensitive individuals and you will have resistant individuals. Why that is in a highly inbred strain like a B6 mouse, we actually don't know. In autism, one could hypothesize that we could have a shift to the left of the sensitivity curves because autism has genes that may impact susceptibility, either through altered metabolism processing or, in fact, more sensitive target sites. But it is more complicated than that because we really believe that there are autisms, not any one type of autism.

This is probably because different populations of kids susceptible to autism have a different pattern of gene expression and different susceptibility markers that contribute to their ultimate susceptibility to autism. So, this becomes a very complicated problem. One way to approach a very complicated gene environment issue is to acknowledge that genetic susceptibility will play off of environmental exposure, and if the timing of those exposures are correct or critically timed, then the prevalence and severity of the developmental disorder will be influenced.

Let's take an example. This is a very simple example. Timothy syndrome is a very rare disorder. It is not a genetic heritable disorder that wipes out hundreds of genes. It is not one that deletes a gene. It is one that actually inserts a missense mutation in the coding region of Cav1.2, changing a G at 406 to an R. The single missense mutation causes the calcium channel for which it codes to inactivate much more slowly than it should. This leads to an abnormal calcium signal. If you look at the kids that suffer from Timothy syndrome, you see that they have a 60 percent autism rate, mental retardation at about 25 percent. I am trying to follow the lines here—21 percent rate of seizure disorder.

The reason that these kids were identified initially was they have long QT. They have an arrhythmatic heart, which contributes to the clinical presentation and, yet, 80 percent are on spectrum; 60 percent are autistic. This is a very rare disorder, but could it lead us to understand gene–environment interaction? Well, it is well known among toxicologists that one of the major targets of Cav1.2 and some of the other L-type calcium channels is mercury, cadmium, and lanthanum. In fact, we use them routinely in the lab to block these channels.

So, this provides a kind of homework, a simple model where you can go in, make a mouse, and then test the hypothesis of a surgical strike on a particular gene and how it might influence genetic susceptibility. We need to look at this as a system. We now know that PTEN highly regulates through its effects on PI-3 kinase-dependent phorylation of PKB, the inactivation kinetics of Cav1.2. PTEN has been another susceptibility marker for autism.

This presents a framework from which we can learn from rare disorders a particular strategy to use to try to understand a complex disease, a rare mutation to understand a complex disease.

Now, let's get a little more complicated in the sense that there have been several neurobiologists who have proposed that one of the sort of patterns you see in autism that may be more generally applied is an imbalance between excitation and inhibition within the developing nervous system. Here now we have several pieces of evidence. Some are more controversial than others, but, nevertheless, they converge on a common defect, which is a more general defect. It is a functional defect in autism, which is a deficiency in GABA-ergic signaling. GABA is the major inhibitory neurotransmitter in the nervous system, and we now have evidence for methylation problems involving the MeCP2 protein, which is associated with a host of genetic outcomes in terms of transcriptional outcomes, but one of the genes that is impacted is the GABA receptor beta 3 subunit.

GABA receptors are also influenced through—have been identified to be involved or at least linked to—autism through linkage studies, and they suggest complex epistatic interactions between, let's say, the GABA receptor alpha 4 and GABA receptor beta 1 genes. Finally, polymorphisms between GABA receptor alpha 1 and GABA receptor beta 3 have also been suggested, as indicated by Pat Levitt. So, how might this imbalance in GABA-ergic transmission be influenced by pesticides?

Well, toxicologists have known for years that one of the major targets of chlorinated hydrocarbons is, in fact, targeting GABA receptors. That is the way in which they work in insects and how they have been proposed to at least have acute toxicological effects in mammals. Then I want to touch on how PCBs might be modifying this.

So, several polychlorinated hydrocarbons of historical importance, in the parentheses on this slide are the dates at which—the years in which they were banned in the United States Lindane is still used for head lice control and scabies. Heptachlor, chlordane, dieldrin, kepone, and toxaphene have all been discontinued, but, in fact, they are extremely stable structures and there are exposures that still occur.

How do they work? They essentially block the pore of GABA receptors. So, they decrease inhibitory neurotransmitters in the central nervous system (CNS). You might say these are of historical importance. Why do we even worry about them? The risk factors that they might contribute are relatively small. Well, it doesn't seem like we have learned about GABA receptors in terms of rationally designed new insecticides. What we find here is that one of the major insecticides currently used in every home is one called fipronil. It is a 4-alkyl-1-phenylpyrazole. In 2000 about 800 tons of it were applied. It goes by several names and we used to think and we used to teach veterinary students that this is an insect selective. It was birationally developed and, therefore, it affected insect receptors and mammalian receptors; a paper back in 2004 did a comparison between the beta 3 expressed GABA receptor and compared it to the insect receptor and essentially showed no selectivity. This again provides a framework of how a GABA-ergic deficiency in autism may play off of an environmental exposure.

Then finally in the broadest sense, we are all exposed to very low levels of persistent organic pollutants. One chemical class is the polychlorinated biphenyls. In a paper that will probably appear next week, there is now evidence that low-level perinatal exposure to these chemicals can, in fact, shift the balance between excitatory and inhibitory currents within the auditory cortex. What I show here is work from Tal Kenet and Mike Merzenich, which shows the normal tonotopic map of a rat at postnatal days 35 through 50, very nice organization and nice gradation from high frequency to low frequency, from blue to red and with a perinatal exposure associated with this imbalance and excitatory over inhibitory current, you see that perinatal exposure to the PCB is called a disruption of the tonotopic map. Now, this is not seen as an overt toxicity in the rat, but certainly if one could imagine this occurs to even a minor extent in children, it would affect language development.

So, I want to finish, that a framework for future studies would be to accept the fact that there are several very complicated genetic susceptibilities in autism and the number and timing of environmental exposures need to be better defined and need to be relevant to the condition. We also need to pay attention to repair mechanisms that may be impaired, such as the DNA methylation.

Here I have given you some examples of how this might work in a real hypothesis-driven research proposal where you might look at competitive and noncompetitive GABA blockers, alterations and self-signaling that may be modified and play off of those mechanisms, such as PCBs, PBDEs, and PCDEs, and then in terms of the framework for future studies, I think we need case-controlled studies so that we can have better comparisons and define subsets, as has already been mentioned.

We need to pay particular attention to immunological susceptibilities because the very genes that I mentioned here play a major role in immune regulation, including Cav1.2. We need molecular, cellular, and *in vivo* models that address mechanistic and behavioral outcomes at low subtoxic exposure levels. I think this is extremely important. We need to better define endpoints and, as will be discussed later in the meeting, nutritional-based models are also extremely important.

So, I thank you.

Dr. Spence: Thank you, Dr. Pessah.

Does anybody have any questions related to this talk?

Dr. Insel.

Dr. Insel: Just as a question, in much of the work that Pat Levitt described, the genetic discoveries are going much more toward the sort of discovery-driven approach without necessarily having a hypothesis about a specific candidate. Are we able to do that in the realm of toxicology? Is there some kind of non-hypothesis-driven exploratory approach that we can take?

Dr. Pessah: I think there are several individuals, some I think will speak later, that one could use simple cell culture models to identify changes in signaling pathways. These are relatively rapid throughput and discovery sort of oriented type approaches where one could, in fact, use individual compounds that are thought to be a potential problem and complex mixtures and identify how specific signaling pathways are altered or some morphological changes are impacted. But that I think is really how do you relate that back to humans, and autism is really the million-dollar—it is a very tough thing to do.

That is why trying to understand some of the susceptibility genes and seeing if, in fact, we already know that they are targets or the pathways are targets may prove to be a bit quicker to try and identify risk factors and interventions.

Dr. Spence: Other questions?

Dr. Schwartz: This is David Schwartz. I was wondering if I could respond to Dr. Insel's question.

I would like Larry Needham to potentially add some information here. There are panels of toxins that one could look at in the serum, blood, hair, and other specimens and apply that in a very broad way to population-based studies. It just hasn't been done. The concept of approaching environmental etiology in an agnostic way, looking at as many toxins and exposures as possible, makes enormous sense, especially in a disease like this.

Larry is in charge of the unit in the CDC that has developed over 150 assays to look at chemicals and toxins in various human specimens.

Dr. Needham: One thing we have talked about—what we have done and I will talk more about it tomorrow—is our division, and Eric Sampson is the director of our division and Henry Falk is the director—what we have done with the NHANES [National Health and Nutrition Examination Survey] survey is look now at about 200-and-some-odd chemicals, but we looked at the general population and, of course, we have talked about folks being on selected populations and doing educated studies and looking for increased amounts or decreased amounts and increases and decreases over time of concentrations of various chemicals.

But that is where we need your help in selecting these chemicals for autism.

Dr. Insel: But we are talking about hundreds, not millions.

Dr. Needham: And also should we be focusing on those chemicals that are only changing in terms of concentration in the environment.

Dr. Schwartz: Dr. Isaac Pessah started off by saying there are roughly 49,000 compounds and 80 percent of them still need to be investigated. We have no way—and the way that we do in genomics now where we have a comprehensive approach, there is nothing comparable here—but we are in the process of developing that. That is the thrust of the environmental biology program. Think about where genetics was 10 years ago in terms of looking at hundreds, not tens of thousands.

Dr. Needham: We need clues like on structural activity relationships and so forth if we can get some clues there.

Dr. Spence: Should we move on and move this discussion to the end of the session?

The last speaker for this session is Dr. Martha Herbert, who is assistant professor of neurology at Harvard Medical School. She is going to give us a talk about the concept of biomarkers.

DEFINING AUTISM: BIOMARKERS AND OTHER RESEARCH TOOLS[6]

Dr. Martha Herbert

Dr. Herbert: I am happy to be here and I want to reiterate that our instructions have been to give a broad and general overview of what we

[6]Throughout Dr. Herbert's presentation, she may refer to slides that can be found online at http://www.iom.edu/?id=42459.

know and what we need to know. I want to comment that the issue of biomarkers is very pertinent for research and it is also very pertinent to those of us in clinical practice.

Why biomarkers in autism? Right now we have no biomarkers for diagnosis. Biomarkers would help in identifying pathophysiological mechanisms. Different biomarkers in different subgroups would be very helpful in multiple regards—identification of treatable features and prediction and tracking of treatment response. Overall, we need to focus on biology and pathophysiology, and there are multiple levels of the biological hierarchy at which we need to measure, and we need to make a more concerted effort to coordinate measures across these levels of the biological hierarchy.

I would like to propose from the point of view of learning more about autism pathophysiology, more than the thimble that Pat Levitt showed in his slide show, that we consider taking a middle-out approach. Bottom up could be genes up and top down could be behavior down. We have been using in autism research a gene–brain behavior model and I think that what we are talking about here is, and everyone has been saying more than that, it is genetically influenced, not just genetically determined and it is a whole organism. It is a systems model where pathogenesis is now gene–environment and epigenetics, and the biology needs to be broken out in details of molecular and cellular mechanisms, tissue and metabolism, altered connectivity and processing to get to the phenotype. It is the mechanisms that yield the phenotype. They may be caused or triggered by the pathogenesis, but it is the mechanisms that lead to what we call autism.

This involves again breaking out the biology in more detail and it is also at the level of pathophysiological mechanisms that will identify biomarkers and also that will identify biomedical treatment targets. We have been talking about involvement of more than the brain in autism and I think it is interesting to go back to the very first paper on autism by Leo Kanner. Kanner commented on somatic symptoms in almost all of his cases, and I have highlighted, in red, eating problems, tonsils, diarrhea and fever, more tonsils and adenoids, frequent vomiting, tonsils, tube feeding, avitaminosis, malnutrition, vomiting, feeding, bronchitis, colds, streptococcus, infection, impetigo.

One child who didn't have any infections, which would be a question of overactive immune system, frequent hospitalizations because of feeding, colds and otitis media, hormonal problems. These were commented on. They did not fit into a model of autism as a behavioral disorder, but it is interesting to note that they were there. They were not life-threatening problems for the most part. It raises the question of what our thresholds are in a complex disorder for taking symptoms into account in the model.

So, I think that what we need is a move again on emphasizing pathophysiology in my area of brain imaging research. We have been looking at this from a cognitive neuroscience point of view, looking at behavior as it is modulated by regional and neurosystems alterations. I think what is happening now is a greater interest in the tissue changes of the brain and looking at the brain based on the physical properties, the receptors, the growth factors, and so forth that may be targeted by a gene and particularly by environmental factors. I think we need to tackle the intersection between pathophysiology and cognitive neuroscience and I think for this we need a programmatic brain–body biomarker linkage.

I went and did some countings of biomarker-pertinent published articles. I went back to the literature and looked at all of the articles that measured biomarkers, starting in the *Journal of Autism and Developmental Disorders* and every paper in that journal and in its predecessor, *Journal of Autism and Childhood Schizophrenia.* In 36 years, there were 78 articles that had to do with biomarkers. You can see in the color coding of the Excel chart that most of them only measured one biomarker and only measured it once in a small cohort without relationship to other kinds of finger typing, whether it be biological or biochemical.

The serotonin is the only one, which was measured multiple times. I had the hypothesis that as genetics became more organized over the years, there would be fewer biomarker studies, and the graph on the bottom left shows the blue line is a decrease in the proportion of biomarker articles as a percentage of total articles that year, a modest increase in genetics, although it should be pointed out that people don't publish their genetic studies in this journal.

A preliminary count of all the articles in PubMed that use autism or autistic in the title yielded 400-some-odd articles out of about 7,000 that talked about biomarkers, with not a lot of repetition. Now, one of the things that is distinctive about our current period is that we are entering into an epoch where it is possible to measure things in new ways. It is possible to measure large numbers of analytes in small quantities of samples and it is also possible due to informatics advances to link measurements across the levels of biological hierarchy.

This may provide fresh and unique opportunities to get a grip on what is going on in autism. Biomarker challenges in autism—critical issues are involved in terms of recognizing what is going on with the phenotypes so that we understand and measure things that are appropriate to the challenge. Many autistic children—not all, but many—have striking variations in their severities, striking good hair days, bad hair days and I will get back to that in a minute.

There are also chronic features, oxidative stress, inflammation, metabolic perturbations. These are ongoing problems. They raise the question of what environmental factors do on an ongoing basis and not just to perturb development. How do they affect neuronal functioning on an ongoing basis? Treatment responsiveness—I will get back to that in a minute. We are seeing some stable improvement following treatment, multisystem involvement. It has been mentioned and, again, whether this impacts the brain primarily parallel or downstream.

The heterogeneity is enormous and that has been mentioned, autisms, leaving the question open of where are the commonalities and final common pathways and how can that question influence our research agenda. Finally, some of the chronic pathophysiological features, such as the inflammation and the oxidative stress, appear nonspecific so that insofar as these are potentially treatment responsive, it is important to remember that what may be treatable may not be specific to autism.

A particular thing that is important to remember is to not characterize autism as a static encephalopathy. There is a paper in press in pediatrics by Andy Zimmerman and others reporting transient marked improvement with fever, children who will start making eye contact and talking during the course of a fever and then it goes away when the fever resolves or somewhat afterward. Some children will have spikes in function sometimes under conditions of stress or emotional stimulation or they will say something quite articulate when they are normally not verbal, demonstrating a neurological capability of performing at that level, but which for some reason is suppressed, which raises the question of whether treatment is removing inhibition or giving skills or both.

Transient improvement on antibiotics has been reported and I will get back to that. Improvement on allergy medications, variability in function related to food, allergen and toxic exposures and also treatment responsiveness, including published reports of loss of diagnosis with recovery documentation studies in process. Overall, this raises the question that this is not just a disorder of neurological development. It is also a disorder of ongoing neuromodulator impact on brain function. This is a slide from Sidney Finegold at UCLA (University of California–Los Angeles) and the veteran's hospital in Los Angeles. He is one of the authors of the study that demonstrated that oral vancomycin could transiently improve symptoms in autism. This was a follow-up study showing nine variants of clostridial bacteria found only in autistic subjects and three variants of clostridia found only in controls.

These abnormal variants of bacteria can deplete vital nutrients, alter metabolism of xenobiotics, and affect immune function. All of this can cause or worsen metabolic stress. This suggests that in order to characterize subgroups and treat these children, we need to go beyond human metabolites and human genome to look at an extended metabolome.

This is a delectable slide of a child standing in his own diarrhea and I don't think the day of autism is complete without looking at this, but the point here is that if you send a stool sample from this child to a clinical laboratory, it would probably come back negative. The measures that allow these bacteria to be identified are done in research labs, and they are not available to help practicing physicians. This is something that needs to change, and this is not just a research question.

There are two reasons why measurements need to be coordinated across levels. It is not just that there is a great deal of variability among results of genetic studies. There is variability in behavior and now that we know that the genetics are not the only thing, we need to confront the variability at multiple levels. We don't know where the commonality is, where it fans in and what is stable across different people with autism. Is it connectivity? Is it more at the tissue metabolism level? Do all children have inflammation in their brains or only some? We don't know this. We need to get systematic about looking for this across all levels.

Also, toxins, infection/immune, genes, and other things function clinically in a vicious cycle. Genetic susceptibility sets up vulnerability to toxins, which impairs immune systems, sets up infection, which alters gene expression and it becomes a self-amplifying vicious-circle feedback loop.

So, conclusions. First, I would propose that metabolism needs to be a core focus in autism. We know that environmental factors perturb metabolism, even at low levels of exposure. We know that some of the same mechanisms and pathways get hit in the metabolic disorders as in inborn errors of metabolism. But the spectrum and intensity of effect differs. We need to learn about multisystem and multilevel impacts. Metabolism is a target for biomedical treatments and also metabolic changes are a final common pathway on which can converge multiple different genetic mutations. What we need to do is to study how environmental perturbation of metabolism, which is not a disease category that we are taught about in medical training at this time, has different patterns and thresholds than inborn errors of metabolism and we need to develop practice parameters around this.

Also what is needed is to develop infrastructural support of the study of metabolism. It is daunting because of the state sensitivity and the sensitivity to handling of sampling. I think we need consensus meetings to identify measures that are less sensitive to these problems. We should consider "omics" and other profiles, develop standard operating procedures, and in particular have a special focus on environmentally responsive metabolism.

As these questions get clarified, we should develop a repository for metabolic samples of many kinds as determined by consensus with multicenter participation and encourage, strongly encourage, participation with contributions from research projects with well-phenotyped subjects.

With regard to brain and metabolism, we know that brain and metabolism are both abnormal in autism and we also know that this is not consistent in its details between subjects. What we need is to learn how metabolism modulates brain and vice-versa. This requires integration and integration requires infrastructure.

I would particularly suggest that in our studies of the brain, we have a much more concerted focus on characterization of brain as a physical organ, characterization of brain tissue. I also propose that we use more high-temporal-resolution brain function measures since the abnormalities in temporal measures, EEG, MEG, at the millisecond level are closer to being indicators of synaptic dysfunction and particularly that since synaptic dysfunction can be metabolically modulated, this is important.

We need, as I have said, systematic metabolic characterization and we need to have an extended metabolome and also extended genome looking at gut microecology and its disruptions.

We also need to have better characterization of change and treatment, which we know are possible. We need better tools to track treatment and change biologically. We should be studying *n*'s of 1, repeated measures to see in the same individuals see what can change. We do not have good measures of change. We can study individuals who are diagnosed over time. We can study children at risk for autism over time, children undergoing treatment over time, and the marked good hair, bad hair days, for example, a child with a fever who improves or other phenomena like that, children who are off fluids and function better.

We need subgrouping to identify mechanisms and to predict treatment response. We do not have published studies showing the separation of groups, such as the illustration that I have shown, which is a separation of Lou Gehrig's disease, Alzheimer's, and Parkinson's, but this kind of work needs to be done in autism.

There is no reason to think that there will be one biomarker for autism. What we need is profiles of vulnerability and treatability. Environmental perturbation of metabolism is widespread in the organism, but its thresholds are different and the reference ranges we use will not pick it up. Autism's sensitive physiology also may mean trouble for the individual, even when labs are within the population normal ranges. So, clinical reference ranges need to be rethought for this kind of complex disorder as part of our process.

Finally, it has been pointed out that the environmental influences on autism suggest treatability and prevention and even though the focus of today's discussion is not on treatment, I think it is very important to understand that what we are talking about is also very much about identification of treatment targets and of treating them.

Thank you.

Dr. Spence: Are there questions for Dr. Herbert?

Dr. Levitt: I just wanted to comment that there are many patients with clear-cut primary genetic disorders like the Fragile X syndrome and Rett syndrome and many, many others, who have good hair days and bad hair days and you have a lot of this kind of variation, even though they have a primary underlying genetic condition.

Dr. Herbert: I would also like to point out that many or most people with some of those primary conditions don't have autism and people with genetic conditions have high vulnerability to environmental perturbation. So, it doesn't exclude an environmental role, even when there is a known genetic factor.

Dr. Levitt: I am not arguing there is no environmental role. I think there is an environmental role and that these people are suspect and subject to these kind of perturbations, but they are particularly—they are profoundly susceptible to these.

Dr. Herbert: That is very telling. That is very interesting. So, this is a question of using controlled clinical settings to study change using systematic systems, biology, and biomarkers.

Dr. Insel: Martha, I really liked your presentation a lot, but could I get you to just expand a little bit for us so that we understand what you mean about some of the studies you would like to see done? For instance, in metabolomics, what would be the tissue and for the microecology, microbiomics, it sounds like you would focus on gut, but can you give—where would you go? What would be the targets for some of these things?

Dr. Herbert: At the moment, I think we haven't explored even blood and urine samples. I am mentioning gut because I wanted to show that this is more than a human metabolism problem. It is also our pathogens or our commensals, but I think we should have—I do not pretend to be the one who can unilaterally dictate what an appropriate profile should be, but I think there is a lot that could be done with blood and urine samples and with a protocol for suitable spinal fluid, even when you don't have a standard study for children who may for one reason or another get a lumbar puncture, they should have available to them a standard operating procedure that can be used in a clinical setting to send samples to a repository.

I really propose consensus meetings to make that decision. I think this is a very complicated area. I have been asked singlehandedly to offer certain organizations the answer to this and I just don't think it is appropriate. I think planning procedures are what we really need to have happen here.

Dr. Spence: Martha, I actually had a question. As a clinician we do get abnormal labs sometimes and I know in metabolism that the state of the child at the time that the lab is drawn is very important. So, do you want to just speak to the challenges of kind of the reliability of some of these biomarkers and standardization?

Dr. Herbert: I think there are some measures that are more state dependent than others. If we are going to organize a repository, we need to get people whose day job and 24/7 specialty is to handle these things, to identify a set of measures, which are most stable. It may be a limited set of measures. There are some measures where if you don't freeze it within 15 minutes and so forth and if you have it at a different time of day, it is very different. But by no means all of them. So, I think the first step would be to get specialists in multiple disciplines to identify measures, which are more and which are less state sensitive, and I think we owe it to the children to really put the effort into doing that.

Dr. Schwartz: You talked about biomarkers as a way of linking pathogenesis to phenotype in terms of mechanisms, but biomarkers could also be used as a way of biologically phenotyping the disease, and in particular this disease strikes me as a disease that is made up of several subtypes. I just wanted your comments and thoughts in terms of using biomarkers as a way of biologically phenotyping autism.

Dr. Herbert: Absolutely. That was on my first slide and I further would expand what I am saying—what I said about that, which is that treatment response measured with biomarkers is an even further way of subtyping. Some of the nutritional treatments that are used in autism in certain settings are relatively low risk and a difference in response to those could be related to a genetically modulated environmentally sensitive set of differences in pathophysiology. That is a great opportunity to learn more about disease mechanisms as a research probe.

Ms. Bono: I just wanted to make a comment that in the blue folder in front of everyone, I made copies of an algorithm that some of the DAN practitioners—that is Defeat Autism Now—use when they get children into their offices and they start subtyping them by blood and urine markers and how they would treat a child based on gut, based on immune problems. It is in the packet. So, if anyone wants to look at that, it is a good start. It is based on about 4,000 or 5,000 children.

Dr. Spence: Great. Thank you.

DISCUSSION

Dr. Spence: I think we should probably move on to the discussion section. I am going to take one minute and try to sum this up. I think Dr. Herbert did a beautiful job of talking about the complexity in the search for biomarkers, even to the point of extending the metabolome beyond humans. So, how complex does that get?

Add to that, Dr. Swedo's description of the heterogeneity of the clinical picture, where there are tens of autisms, maybe twenties, maybe hundreds. Dr. Levitt's description of the multiple ways in which our 30,000 genes can contribute to this disorder and then Dr. Pessah's description about toxicology and 53,000 known toxicants. So, if you do the math, I think this is a real challenge for us, but I think that we have laid out some of the issues in framing this discussion as to what we actually need to go forward. The complexity is daunting, but I think we need to talk about how to overcome that. One thing that I would start with is by posing a question both to our session speakers and also to the table: Do we really need to understand that middle approach, that pathophysiology? Do we need to get past Dr. Levitt's thimble in order to get going on this or can it be attacked in multiple levels?

Can the geneticist be working on genetics and the metabolomics people working on metabolome? So, that is where I am going to start.

Dr. Akil: I feel like we cannot have parallel play because of the very nature of the beast. We cannot have genetics working over here and toxicologists and immunologists and metabolomics. The main reason I see that is the heterogeneity that everybody talks about. What I am not getting out of the discussion is, what are the variables along which you would separate or segregate? Do you wait and let things be self-segregating? For example, do you do huge numbers of subjects with a huge number of toxicology screens and then let that self-aggregate and say in those tens of thousands of autistic kids, there are five patterns and then plus a whole bunch—where we can't figure out anything.

Do you do the same in genetics or do you intersect or do you do anything a priori behaviorally, like so when you come and say there is this gene or that gene, this calcium channel or this tyrosine kinase receptor, is there anything we can hang our hat on from a clinical point of view? So, I think while it is good to talk about all these approaches, the question is exactly how do you intelligently bring them together with statistical analysis and multilevel analysis and so on. I mean, I think that is what we need to grapple with.

Participant: We are portraying this as an intractable problem and I think there is some preliminary parsing that many people have already done. There is a difference between boys and girls. There are kids who have GI (gastrointestinal) disturbances and kids who don't. There are kids who look abnormal from the get-go and there are other children who clearly are very different, later in childhood. So, I think there are already some areas where you can begin to break this down to some extent.

Dr. Swedo: I would just echo that and say that clinically I think one of the things that we have spent a lot of time on was coming to diagnostic agreement on the behavioral characteristics that make up autism. I mean, if you look at the research papers, it has been largely aimed at better and better and earlier and earlier diagnosis of the social and communications deficits. There has been almost no further look at the mind or the brain within the body.

One of the things that is already underway is major phenotyping efforts to begin to do what you have suggested. Can we find those few kids as well as going after the 10,000 at the same time?

Dr. Akil: Does it map the phenotyping map onto anything else that you can identify?

Dr. Swedo: Exactly, but I think the way we set it up so far is that we were looking just at the two characteristics that would put them into that diagnostic group and couldn't begin to stratify the way Pat was suggesting we would need to for better genetic power.

Dr. Leshner: Can I ask a similar question? I had actually the same question that Huda Akil has just asked. That is, is there a group with some common credibility or standing, who, in fact, could do the level of detailed phenotyping that you would need to—or at least to be able to parse the symptomology into agreeable form so that you could look at individual clusters of symptoms or look at subtypes or whatever in a way that might get us off the dime a little bit more easily? But who does that?

Dr. Swedo: I will speak for my colleagues at the NIH and the CDC, but they might want to chime in as well. The Autism Phenome Project, that phenotyping effort, was a short-term goal of the IACC (Interagency Autism Coordinating Committee) research matrix and efforts are already underway. Our intramural research group is collaborating with the M.I.N.D. Institute on the pilot project for in-depth phenotyping, biological metabolomics, neuroimaging, all of the variables that we can—the purpose of the first hundred children to be evaluated is to determine what is feasible, how reliable are things among different sites.

The autism centers of excellence RFA was actually written for an impressive number of common measures that will include medical history, environmental exposures, and other things. Every child evaluated within one of the new autism centers of excellence will have those common measures done and those data will be entered into the National Database for Autism Research, NDAR, in an effort to very rapidly get large enough populations to start doing the phenotyping.

Dr. Levitt: I just want to point out one other thing. The domains that we utilize to both diagnose—the functional domains that we use to diagnose and then to phenotypically subcategorize—are those neurobiological domains by definition that are so heterogeneous within the typical population. We are not talking about measuring grip strength here. Social behavior is by definition broadly heterogeneous within that normal distribution, and communication and language development from model systems and birds, all the way up through humans depends upon social behavior, social communication, and social interaction.

So, it is a problem in terms of division and definition because of the very components of brain function that we are focusing on, and you see these domains disrupted in the broad spectrum of psychiatric disorders and it is equally difficult to phenotype in those as it is in autism spectrum disorders.

Dr. Leshner: Not to perseverate too much on this, but I am one of the people who believe that some progress has been delayed in other brain and mental disorder conditions by overlumping and an overfocus on diagnostic categories and, therefore, focusing on symptom clusters or particularly disabilities, however you want to cast it, may be another route.

Dr. Levitt: I agree with you completely. So, the characterization and the divisions and the stratifications are extremely important. I am just saying that the challenge is because of what we are given, the domains that are most difficult in terms of human behavior and function to characterize precisely—and from a practical perspective, if you go around centers around the country, if you want to have an integrated approach, whether you are going to do a metabolomic approach or to integrate that, you need people who are really good, as good and as precise as the quality control we now insist upon in gene sequencing to characterize the populations and it is—I don't know. Maybe I am—but it is extremely difficult to find those individuals; you need clinical psychologists and others who can do this well or who want to do it because it is an enormous task.

If you go around the country and say have you been able to hire a clinical psychologist to work with your geneticist—because I can tell you that the genetics studies from my perspective, the non-syndromic genetic studies, are tainted simply because the populations that have been included are not well characterized. So, this is a real gap. This is a people power gap, just like nurses are a people power gap in hospitals in terms of medical care.

Mr. Blaxill: Alan used the word "standing" and I think one of the things that is interesting about the problem clinically is that there are a lot of clinicians out there working the problem on a different model, on an environmentally based model, on a gut model, on an inflammation model and oxidative stress model, which are sort of the things that are beginning to come into the discussion more frequently. There are a lot of people out there clinically who have been working the problem, but they tend not to have standing because they have been outside the mainstream of where the science has been.

So, I think one of the things we need is to break down some of the cultural barriers to, oh, gosh, those people are kookie. They are doing crazy stuff. Some of them are, but a lot of them are actually conscientious, good clinicians, trying to treat sick kids and we ought to take that community more seriously than we do. They have a lot of data and they don't have any resources and if we spent some time, you know, learning from recovery, learning from treatment response, working with some hypotheses about inflammation and toxicology. There is a lot that could accelerate in terms of getting data faster, testing hypotheses faster and would serve to break down a lot of the divisions that have emerged emotionally about the whole issue. So, I think there is a big group out there that doesn't have standing and we ought to reach out to them maybe a bit more than we do.

Dr. Leshner: I would like to make one comment and then I will shut up, I promise.

I think that the essence of real translational research, though, is to, in fact, be listening to the clinical experience, have data inform the scientific agenda and so I think you are right. The mechanisms for doing that, our institute directors have to figure that out, not I, are really complicated and difficult because sorting through the inappropriate stuff to get to the appropriate stuff could be very difficult. But your point is very well made.

Ms. Redwood: I would like to follow up on what Mr. Blaxill said from an advocacy perspective and what Laura presented as well. There is a sense of urgency here. I am concerned or my belief is that we don't really have to understand mechanisms to be able to intervene in a meaningful way with these children. They are very sick. I think the slide that Martha Herbert put up with regard to their gut disorders, there are things that we can do now if we focus on that *n* of 1 that Martha mentioned and work on trying to help these children medically. They have several medical problems.

I know my son, for example, had very low cholesterol levels. They were in the nineties. He wasn't digesting his food. We treated him with Creon, which is a prescription digestive enzyme. He gained 14 pounds in one year and because his brain was able to finally get nutrition, he improved functionally. He had very elevated levels of serum B_{12}, but he also had methamolonic acid in his urine. He had a functional B_{12} deficiency. Treating him with B_{12} resulted in marked improvement. So, there are things we can do now.

I would sort of argue with Martha not to focus on the brain, but to actually focus on the child with an *n* of 1 and document these medical abnormalities and treat them. We can do that now without necessarily having to know all the mechanisms involved.

Dr. Newschaffer: I just wanted to add something. I think that the emphasis on etiologic heterogeneity and subtyping autism cases is critical and a number of folks have touched on that. What I haven't touched on this morning and it is a little bit out of my wheelhouse—maybe Dr. Levitt can comment on this—is the idea that we might also want to look a little bit at continuous outcomes related to the autism spectrum, the idea of endophenotypes or looking at continuous measures of social cognition in populations that perhaps include individuals affected with autism spectrum disorders, but also broader samples. Continuous endpoints are favorable to study in a number of different study designs, looking at—that even include examinations of gene–environmental interaction.

So, while the etiologic heterogeneity and subtyping I think is paramount and that became clear, too, we also might want to think a little bit about continuous endpoints broadly distributed in the population and how that can inform what we know about gene–environment interactions on related behaviors.

Dr. Herbert: A couple of things. I didn't mean to say we have to start with the brain. If I said that, it was a complete misunderstanding. I mean, obviously, I work in brain, but I think this is a whole-body condition.

We are trained as scientists to really like precision and definition and careful definition. We are dealing with a situation—the figure I have heard of the number of chemicals that we don't know very much about is much higher than 53,000 now. That was a 1992 number. These come in different combinations at different times and they pass through a certain number of final common pathways in our bodies, which our physiological capability is to handle them. But beyond a certain point, it is not going to be that precise. There is going to be continuous distribution not only because things are normally continuously distributed in the population, but because the injuries are continuously distributed and they are not necessarily going to parse out quite uniquely. So, I think that we are going to have to do some very existential reflection on what it means to design a study when it is never going to come out neat and that we need to proceed anyway with taking care of people with the parts that we do know how to handle even while we don't have a grip on many of the other parts.

That is just the nature of the beast. That is the nature of gene–environment interaction in a situation where we did not think through ahead of time the impact of the individual or combined exposures that we are dealing with.

Dr. James: I think there may be a unique opportunity in treatment trials. So far the placebo-controlled, double-blind studies have been disappointing. We basically don't see differences between the control and the autistic group, but I think it is important to look beyond the mean. I think within the mean there are responders and they are not characterized. If we could look at those that do respond and characterize them, this would be a very productive way to look at subpopulations and be able to get perhaps to more individualized approaches to treatment because I do believe that within the mean, there are responders and we should not neglect them. I think there is a huge opportunity to look at subtypes and individualize approaches to their treatment.

Dr. Swedo: That could be actually very useful in other studies where they have done that. For example, there is a classic study in obsessive-compulsive disorder where responders to behavioral therapy and responders to medicine have the same types of changes on their PET scans. So, I think the response is key. One of the questions, though, is how do you look at the responders to placebo. I mean, that is what I struggle with in our placebo-controlled trials. I agree with you completely that you want to break it down to responders and nonresponders and see what those differences are. But if all they have received is a sugar pill and they respond, how would you go after them? Have to look at those two subgroups very carefully and look at what is different about the autistic children who did respond and hopefully might find something there.

Equally important, I think, are the negative responders because within the spectrum, you will have children who absolutely have clear responses and then others who go the other way, who regress. I think they are also equally important to characterize individually, again, working toward more individualized approaches to their treatment.

Dr. Insel: This is very helpful. I am kind of listening to this discussion to get ideas about where the next generation of studies might be and what I am hearing is that part of what makes it such a difficult problem is it is an equation with two variables and we don't really have a good handle on either one. One option people are saying is that you could do within subject designs so that you don't have to worry about genetic variation at all and just look at how changing environmental factors alters outcomes.

There is the real attraction in doing that in kids who may have responded. Would it also be useful if we had a repository of twins, where you could look at discordant twins or, again, you would take genes out of the equation or at least genetic sequence out of the equation and then be able to ask why does one child get the disease or why does one kid become more severely affected and another one not? Has that been done? Is that an option? Is that something we should chase?

Dr. Lipkin: Tom, when we started putting together the AGRE (Autism Genetic Resource Exchange) database, which goes back to the mid-nineties, mid- to late nineties and I was the chair of the first Scientific Advisory Board, I argued for collection of 14 monozygotic (MZ) twins discordant for disease. At that point it was a small fraction of what we found, but there was no interest in proceeding with that. I would imagine we can still find these kids and there must be many more at this point who have been identified. It would be a useful thing to try to do.

Dr. Insel: I think Dr. Hu has actually done some of this, using transcriptomics to—do you want to say something about that?

Participant: My name is Valerie Hu and last year we published a small study on monozygotic twins. That is identical twins that were discordant for autism and we found differential expression. That means turning on or off of a number of genes, the majority of which or at least half of which had no neurological functions in lymphoblastoid cell lines from the AGRE repository. We have since continued the study with case controls, sib pairs, and we found additional genes that play a role. They are pointing toward cholesterol metabolism and androgen biosynthesis. This is all new.

We have also just started a pilot study looking at the epigenetic effects in the same monozygotic twin pairs who we studied before and we are getting confirmatory results supporting the same genes, some of the same genes that we have identified by expression analysis and others. If you do a pathway analysis and look at some of the canonical pathways that are established biochemical and signaling pathways that are implicated by both the genes that we have identified by gene expression analysis as well as by the preliminary methylation analysis, they really converge and they have pointed to some very interesting pathways, some of which are really surprising where they confirmed the involvement of the steroid receptor activity signaling, as well as what surprised me, Type 2 diabetes signaling and insulin signaling. So, that might be a tie-in to some of the additional systemic problems.

Dr. Lipkin: This was done with cell lines rather than primary tissue?

Participant: Lymphoblastoid cell lines. So, you have to consider the caveat in using those.

Dr. Lipkin: The notion at the time that we started when we proposed this was to collect cells, so that the RNA would be there available in PAX gene or Tempest Tube or so forth. That would be—it would perhaps even give you cleaner data. But I am very excited about what you have just described.

Dr. Insel: What I am trying to hear from this group is where are the opportunities like this where you could control one of the two variables, either hold onto the genetics and take that out of the equation or hold onto the environment and take that out of the equation. I think we need some sense of what that range of possibilities would be because you are not going to solve this equation with those variables flipping around on you.

Participant: What we are trying to do is to take a systems biology approach and Dr. Herbert referred to this earlier in terms of the approach that she would recommend. We are trying to pull in not just the genomics, but also the metabolomics plus the epigenetics and we are trying to construct a neuronal cell model so that we can in vitro—well, one of the processes that seems to be constantly implicated not just by my studies, but many other people's studies, from the M.I.N.D. Institute by Daniel Geschwind's group, many others, is that of axon guidance or neuroextension, neuronal cells going out and finding their targets.

So, we want to establish a neuronal in vitro cell model so that we can test the impact of various stressors that are both environmental as well as biologic stressors, such as elevated testosterone, for example, which is implicated in our study. I think an integrated approach is really necessary.

Dr. Spence: Thank you, Dr. Hu.

Dr. Susser.

Dr. Susser: I wanted to agree but also underscore Tom Insel's point. I think that when we talk about complexity, what we are really trying to do is talk about it so that we could simplify things and, you know, one approach is MZ twins, so it simplifies one part of it. The other approach is to look for situations where people have common environmental exposure, which you implied in your comment, too. There are many such situations that we can identify and we can exploit.

My view on this is that it becomes infinitely complex if we say that we have to know everything about everything and the way that we will get to identify causes, at least, is by thinking about designs, exemplified by the MZ twin design, but that would just be the controlled in genetic practice. Think about also where you control environmental factors and look for genetic effects. There are many such and I will just throw a couple out so you know there are possibilities.

You can look at groups that have been exposed to congenital rubella, to rubella in utero. You can look at groups that have been exposed to different kinds of toxic poisoning. You could look at groups that have very old fathers, for example. There are many possibilities in this range of looking for designs that simplify the problem.

Dr. Akil: I think it was Susan earlier who said comparisons across different cultures, right? The Institute of Medicine has sort of a big interest in global outreach. I sort of don't want this idea to go away. That is going in like the very other, very big direction instead of controlling everything within like small number of pairs of identical twins, asking the very broad question, which I think is again a very different approach.

Two other little points. One is that in these twin studies, it would still be wonderful if we had as much genetic information as possible, as well as gene expression profiling. If we could afford to sequence it, I would sequence it, but short of that, it is not that expensive to get, because different pairs of twins may turn out to be different and having that information would be very helpful.

Finally, I hope we keep in mind protective as well as vulnerability factors in all of our thinking.

Dr. Spence: That is an excellent point. I think we are going to finish up and we have one more question from Dr. Choi.

Dr. Choi: These equation-solving efforts seem to me to be very important, but their ultimate impact to my way of thinking likely does rest on the platform of understanding the phenotype. I mean, essentially there is potentially a whole series of equations out there. So, imagine a worst-case setting where you have 20 monozygotic twins and each one has a different autism. So that it is very difficult to pull the signal out of noise there.

So, I go back to the need to really get at phenotyping and I am sort of struck as an outsider to this field by what sounds to me like a relative need for a bit of catch-up ball here in getting at the phenotype; understanding that clinical phenotype when you are dealing with cognitive and behavioral disorders is very challenging. It does seem like the biological phenotyping effort is really lagging behind that effort in several other fields. I am really struck when you say that even blood and urine haven't been thoroughly examined, given the rich tradition of looking at those fluids, particularly urine in other cognitive and medical disorders.

So, it seems like that is something that ought to be a full-court press. I hope that is part of the national effort.

Dr. Leshner: We will add that to the agenda.

As keeper of the clock, I am going to call this session to a close. I want to thank Dr. Spence and the speakers, who have done a wonderful job.

I do want to comment on what a wonderful array of people are in this room. This speaks, I think, to the commitment of many, many people. We have institute directors from NIH, the deputy director of the National Science Foundation, the basic science agency and an array of wonderful scientists, both clinical and basic. So, I think we are well on our way.

Session II
Lessons Learned from Other Disorders: "Standards of Evidence"

Dr. Leshner: I think the first session got us off to a wonderful start. Again, we appreciate the speakers tremendously. The audience isn't too bad either.

I would now like to introduce Dr. David Schwartz, the director of the National Institute of Environmental Health Sciences, who will chair the next session.

Dr. Schwartz: Thank you, Alan. It is a pleasure to be here. Good morning. Welcome to Session II. Lessons learned from other disorders: Standards of evidence.

Over the past couple of years, it has been a real pleasure for me to get to know the autism community, both the advocates and the scientists in the community. What has become abundantly clear to me is that this is a real challenge to our institute to try to help understand what is causing this disease and also how the environment can contribute to the understanding of the pathogenesis of this disorder as well as the phenotyping of the disorder.

This session is set up as a way of taking some environmental diseases, diseases where we have uncovered causes in the environment that are related to complex diseases, identified the causes, and identifying the causes has led to a much richer understanding of the pathogenesis, genetics, and also ways to prevent the disease processes.

So, to start this off, Phil Landrigan is going to give the first talk on environmental toxicants and neurodevelopment. Phil Landrigan is a pediatrician at Mt. Sinai School of Medicine. He is also an epidemiologist with a long history in environmental sciences and he is chair of the Department of Community and Preventive Medicine.

ENVIRONMENTAL TOXICANTS AND NEURODEVELOPMENT[7]

Dr. Philip Landrigan

Dr. Landrigan: Let me begin by thanking the organizers of this conference. This is a very important gathering that brings together people from the community of concerned parents, the autism research community, the genetics community, the public health community, and the clinical pediatric community. The only way we are going to make progress against a complex and multifactorial disease such as autism is to have people of these diverse backgrounds share their insights and talk across their boundaries, as we are doing here today. I salute the IOM organizers, and I thank them for the extraordinary preparatory work that I know they have selflessly undertaken to make this day possible.

The two central questions before us today are: (1) how can we accelerate the discovery of new knowledge about the preventable environmental causes of autism; and (2) how can we effectively translate these discoveries to the clinic and to the community to improve treatments and to strengthen the prevention of autism.

I will approach these two questions by presenting two case studies—the cases of lead and of the organophosphate (OP) pesticides. Our work over the past several decades on lead and OPs has taught us a great deal about how chemicals can injure the developing brain. And additionally this work has taught us much about how to translate science to treatment and to prevention. I suspect that there may be many parallels in these examples that are relevant to the case of autism. I shall start with lead.

When the medical and scientific communities first came to recognize lead poisoning more than 2,000 years ago, the condition was thought to be an occupational disease that principally affected adults. Lead poisoning in ancient times was seen principally in miners, smeltermen, painters, and potters.

Starting in the late Middle Ages and through the industrial revolution, our species started spewing lead widely into the environment. That environmental dissemination accelerated sharply in the 20th century with the addition of lead to gasoline. As a result, many people in addition to workers came to be exposed to lead.

---

[7]Throughout Dr. Landrigan's presentation, he may refer to slides that can be found online at http://www.iom.edu/?id=42461.

Beginning in the early 1900s, physicians began to realize that lead poisoning could affect children. Until then, kids were thought to be immune to lead. The initial discovery of childhood lead poisoning was made by a clinician and an epidemiologist working together in Australia, Dr. Gibson and Dr. Turner, who were confronted clinically with a group of children who had GI problems, headaches, coma, and convulsions. Some of these children died. The disease was thought initially to be infectious, some kind of encephalitis. It was only through patient detective work over 14 years that Gibson and Turner came to realize that these children were, in fact, lead poisoned and that their poisoning had resulted from ingesting lead paint chips and contaminated dust on their verandas.

The medical community came to learn additionally over succeeding decades that lead poisoning is a disease that can cause damage to children even in the absence of clinical symptoms. Prior to the 1940s, lead poisoning was thought to be a disease that either killed a child or from which the child recovered. But in 1943, Dr. Randolph Byers, a pioneering pediatric neurologist in Boston, realized that children who had suffered from lead poisoning, and who were thought to have fully recovered, could come back into the hospital a number of years later with behavioral problems. The event that triggered this clinical observation was an episode in which one of Dr. Byers' former lead patients, a boy who had previously been a docile child, stabbed a teacher with a pair of scissors. That dramatic story prompted a study which showed that 19 of 20 of previously lead-poisoned children, who were thought to have recovered, had persistent hyperaggressive behavior.

Dr. Byers' work paved the way for work that our group at CDC did in the 1970s and that Herb Needleman's group in Boston and then in Pittsburgh did in the late seventies and into the eighties. These studies showed that asymptomatic children who were exposed to lead—lead from a smelter in El Paso and lead from paint in Boston—could have a decreased IQ, shortening of attention span, and problems in school. These effects were dose related and were more severe in the more heavily exposed children. They occurred entirely in the absence of clinical symptoms of lead poisoning.

These studies introduced the notion of subclinical toxicity, the concept that there is a continuum of toxicity in which the clinically apparent effects of lead or other toxins have their subclinical counterparts. We came to recognize that lead could have subclinical manifestations, defined as damage that is not clinically obvious, but that is quite real and very readily demonstrable on special testing such as testing of intelligence, attention span, or impulsivity.

We have come to realize further that lead affects many brain functions in addition to intelligence. Behavior is another target of lead, and children with subclinical lead poisoning fail in school. They fail at life. They are dyslexic. They drop out. They are incarcerated.

We have been able to recognize these causal connections between lead and brain injury because we have reliable exposure measures—biological markers—namely, the measurement of lead in blood, and more recently the measurement of lead in bone. In undertaking epidemiologic studies of environmental exposures, it is incredibly important to have stable biomarkers of exposure.

Three further lessons emerge from the case study of lead.

The first is recognition that when a neurotoxic chemical is widely dispersed in society as was lead in the years when we allowed it to be added to gasoline, subclinical toxicity is likely also to be widespread and can affect entire societies. In the 1970s, we were putting more than 100,000 pounds of lead into gasoline each year and then spewing all of that lead out into the environment through the tailpipes of cars. The result is that the population mean blood-lead level among American children in the 1970s was almost 20 micrograms per deciliter, a level that today would be considered dangerously high. It is clear in retrospect that subclinical neurotoxicity was widespread and that IQ was diminished across virtually the entire U.S. population. Moreover, if the mean IQ in those years was reduced by just 5 percent, as it almost certainly was, the result would have been a reduction of more than 50 percent in the number of gifted children and a corresponding increase of more than 50 percent of the number of kids who are going to have problems.

Second, we have learned that not all individuals are equally sensitive to neurotoxic chemicals such as lead. Genetic and physiological differences convey sharp differences in vulnerability. Accordingly, we have begun since the decoding of the human genome to explore gene–environment interactions that influence susceptibility to lead. I suspect that gene–environment interactions may be important also in the genesis of autism, and that some of the lessons that we are learning about lead will be relevant to the understanding of autism.

The third, critically important lesson that emerges from the lead case study is that neurotoxicity which is caused by a toxic chemical in the environment can be prevented. In 1976 as a consequence of our epidemiologic studies, and following additional studies of lead neurotoxicity that were supported by NIEHS in Cincinnati, Australia, and the former Yugoslavia, EPA made the decision to phase lead out of gasoline over a multiyear period. It was predicted by EPA that there would be a very modest decline in the population average blood-lead level, perhaps a 1- or a 2-microgram decline.

What happened in reality was that there was a 50 percent decline in the average blood-lead level that paralleled incredibly closely the decline of the content of lead in gasoline between 1976 and 1980. This decline has continued to the present day, so that the average blood-lead level now in the USA is less than 2 micrograms per deciliter. In other words, we have achieved a better than 90 percent reduction in blood-lead levels in this country as a consequence of our scientific discovery of the developmental neurotoxicity of lead.

Good science has driven this process at every step. But science alone was not enough to achieve prevention. Prevention required partnerships among scientists, regulators, elected officials, pediatricians, concerned parents, and society in general. The lessons for autism are clear.

So, the question for those of us who care about public health is what can we do to speed this up.

I think my second case study, the study of the neurotoxicity of the OP pesticides, is illustrative of the acceleration in the pace of discovery that can be achieved when appropriate resources are directed at a problem.

Studies of the developmental and pediatric neurotoxicity of OP pesticides was triggered by a 1993 report from the National Academy of Sciences, entitled *Pesticides in the Diets of Infants and Children.* This report highlighted the unique exposures and the special vulnerability of children to pesticides. Its findings paved the way for passage in 1996 of the Food Quality Protection Act (FQPA), the principal federal law that governs the use of pesticides in agriculture.

That law spoke for the first time in this nation in the language of law and policy about the importance of affording children special protections in law and regulation against neurotoxic chemicals. The passage of FQPA led to an outpouring of investment in children's environmental health. This research was needed to fulfill the requirements of FQPA, which called for child-focused studies. It led to the creation of a national network of Children's Environmental Health and Disease Prevention

Research Centers, the National Children's Study, Pediatric Environmental Health Specialty Units supported by CDC, creation of the Office of Children's Environmental Health Protection at EPA, and a whole outpouring of investment in a field that had previously been very seriously neglected. The next figure is a curve showing the increase in funding over the past decade in children's environmental health.

A specific consequence of this increased investment in children's environmental health is that in the study of organophosphate pesticides and their effects on the developing brain, we have made great progress, and we have done so at a far more rapid pace than in the case of lead.

This rapid progress in understanding the developmental neurotoxicity of the OP insecticides began with Ted Slotkin's work at Duke in which he showed that exposure of newborn rodents to organophosphates could cause anatomical problems in the brain, reductions in the number of cells, and behavioral problems. These deficits were fixed and not reversible.

This work was followed by studies showing that the genetic variation in the enzyme paraoxonase (PON), which is centrally involved in metabolism of OP pesticides, could profoundly influence susceptibil-ity, a clear example of gene–environment interaction.

Those studies were followed by the development of exposure assessment strategies, in which scientists learned how to measure organophosphates in body fluids, especially in urine. We learned also how to measure organophosphates in air and dust in homes, where they had been applied to control cockroaches in urban apartments. We found that these allegedly short-lived chemicals actually had a residence time in apartments that could be measured in weeks or months.

Then we moved into the realm of clinical epidemiology. In this effort, our group at the Mt. Sinai School of Medicine and our colleagues at Columbia University and at University of California–Berkeley recruited populations of mothers and their children who were followed prospectively. We recruited the moms when they were still pregnant, and we assessed the moms' exposure to OP pesticides during pregnancy by measuring levels of OP metabolites in maternal urine.

A critically important finding was that babies who were exposed during pregnancy to OPs had small head circumference at birth, a measure of delayed brain growth during the 9 months of pregnancy. This effect was especially striking in babies born to mothers who had low expression levels of PON. These babies had developmental delays. They had cognitive defects. They had increased risk of attention deficit hyperactivity disorder (ADHD). And most important to our discussion here today, the babies exposed to OP pesticides during pregnancy appear to have an increased incidence of pervasive developmental disorder, which of course is a component of the autistic spectrum.

As a consequence of these findings, EPA banned residential use of the two most widely used OP pesticides—chlorpyrifos and diazinon. Reduction in the frequency of impaired babies was documented within months to result from this ban.

In summary, these two case studies teach us the following lessons:

- Chemicals in the environment can injure the human brain.
- Children are especially vulnerable to brain injury caused by chemicals, and this vulnerability is generally greatest during the 9 months of pregnancy and in the earliest years of life.
- The brain injury caused in children by chemicals is sometimes symptomatic, but more often produces a range of abnormalities that impair function and that can be detected only through special testing.
- Chemicals can cause syndromes in children, such as ADHD and PDD.
- Chemically induced injury to the developing brain can be prevented by the application of scientific discovery.
- The pace of scientific discovery can be dramatically accelerated by focused investment in research.

Conclusion: Where do we go from here? How do we apply the lessons learned from study of the neurotoxicity of lead and OP insecticides to better understand, treat, and prevent autism?

I will argue that an overarching need is to build on the investments our society has made in the past decade and to continue to support research in children's environmental health. Without continuing support for research, there will be no discovery. And if there is no discovery, there will be no new treatments and no new prevention.

I identify four specific needs:

1. We must as a nation continue to support Centers of Excellence in Children's Environmental Health and Disease Prevention Research. There exists a strong and highly productive national centers program. It is under review. Review is a good thing. As a

result of this review, there will be some change in the composition of those centers. However, the essential need is to continue to sustain multidisciplinary centers in children's environmental health, and most especially to sustain the prospective birth cohort studies within those centers, which are the jewel in the crown.

2. We must support a large national prospective birth cohort study of American children such as the National Children's Study. The National Children's Study will follow 100,000 children, a statistically representative sample of all children born in the United States over a 21-year period from conception to adulthood. The goal is to identify the preventable environmental causes of autism and other diseases of children and then to apply those scientific findings to create a national blueprint for treatment and prevention.

   Given the current prevalence of autism in the U.S., a study of 100,000 children will give us almost 1,000 children with autism and 99,000 controls. Moreover, because the study will collect data on hundreds of environmental exposures (to be measured by CDC) and on the individual genetic susceptibility of each child in the study, it will provide us an unparalleled opportunity to examine interactions between genome and environment in child development. There will be no better opportunity in our lifetimes to discover the preventable environmental causes of autism.

3. We need training programs that increase the national workforce in environmental pediatrics. Today far too few pediatricians have more than minimal understanding of the pervasive influence of the environment on child health.

4. We must improve the testing of chemicals for potential toxicity, especially developmental neurotoxicity. We must end the current situation of deliberate ignorance, in which we produce new chemicals, disseminate them into the environment, but fail to test them for potential toxicity.

Today there are over 80,000 chemicals in commerce. Approximately 3,000 of these chemicals are classed as high-production-volume (HPV) chemicals. Fewer than 20 percent of HPV chemicals have been tested for their potential capacity to injure the developing brain. This is an untenable situation.

Thank you.

Dr. Schwartz: Thank you very much.

There is time for one or two questions if anyone around the table has a question.

Mr. Blaxill: Phil, could you comment a little bit on the institutional response to the lead problems and some of the resistance that the science faced?

Dr. Landrigan: Well, yes. There was huge resistance. The problem is that lead was a very profitable chemical in the 1970s and there was a huge lead lobby that did their best to discount every scientific finding that was made in those years in public forums and in private meetings. The lead industry did their best to pillory Herb Needleman, whose picture I showed in the middle of the slide. They had a couple of scientists who were in their pay, although who didn't acknowledge that they were in the industry's pay until later, came forward and charged Dr. Needleman with scientific fraud. His case was hung up for 4 years at the NIH, while that terribly painful process was cranked through. He was eventually completely vindicated and has won a whole series of prestigious awards since that time, but there was great resistance to learning the results of the research or to translating those research results into public policy.

I think some of the reasons today that we have 80,000 chemicals in commerce of which fewer than 20 percent have been properly tested reflect the same legacy of special interests not caring to know about the toxicity of chemicals. I honestly think as a society, we need to get beyond that. We are flying blind if we allow kids to continue to be exposed to chemicals of untested toxicity. It is not a political issue. It sometimes gets portrayed as one, but it is not.

What it is, it is an issue of protecting kids and I think it is an issue that people all across the political spectrum in this town should get together and say we really need to do something about this. We need to test the chemicals. We need to be examining the children. We need to be doing good research that leads to prevention.

Dr. Slotkin: Just to add onto that, the same thing that happened with the lead story is also true even today with the pesticides and particularly the organophosphates, where scientific papers are being fought through letter-writing campaigns and whispering campaigns done by scientists and others in the pay of the chemical companies.

Dr. Schwartz: Thank you, Phil.

We should move on to the next speaker so we have enough time for the discussion.

Ezra Susser is the next speaker. Dr. Susser is a psychiatrist and an epidemiologist. He is chair of the Department of Epidemiology at Columbia University in the Mailman School of Public Health.

PRENATAL STARVATION AND SCHIZOPHRENIA[8]

Dr. Ezra Susser

Dr. Susser: I am going to talk about an example which is less advanced. It is an example where we have established a connection, some kind of link between an early prenatal exposure, prenatal famine, and the emergence of a disorder decades later, which is schizophrenia. But we don't yet know what the causal pathway is that accounts for this link. Learning that would lead us toward prevention and intervention. So, that is where we stand.

It was hard to get to this point. I am going to talk about how we got there and then what we do next. Just a few words first about what the challenges are of establishing a relation between early prenatal exposure of any kind and then neurodevelopment outcome, and many of these pertain to autism, too. The problem with the time between conception to birth is that development is very rapid and we only have indirect ways of assessing what is happening to the fetus. What we usually do actually is measure what is happening to the mother. We use that indirectly as a window on what is happening to the fetus.

So, that is one of the challenges. Another challenge that we have in this area, which pertains to autism as well as schizophrenia, is that we need very large numbers because they are not common diseases and we need to do very labor-intensive assessments to establish good diagnoses. That is extremely difficult to do. It is hard to assess hundreds of thousands of people as to who has autism, who has schizophrenia, and so forth. It is even more difficult for schizophrenia because you may have to wait for 30 years or more after birth before you can make the diagnosis. It is quite challenging to do this.

[8]Throughout Dr. Susser's presentation, he may refer to slides that can be found online at http://www.iom.edu/?id=42462.

But there are a number of examples where we have been able to overcome this challenge. I am going to talk about them. This begins with a study that we did based on the Dutch hunger winter that was at the end of World War II in Holland. It was due to a Nazi blockade of occupied Holland in the last part of World War II. We focused on the people who were conceived at the very height of that famine.

These were people who were born from October 15 to December 31 of 1945 in this region of Holland. The picture there is just a picture of the food ration that was received by the whole population around the time that these children were conceived. It was very meager.

What you see in the two graphs there, in the top graph, it shows the birth rate in this part of Holland across a 3-year period and you see that there is a very dramatic drop in fertility, which follows the drop in food supply around the time of conception. The blue shaded area in the graph marks the group that we identified as exposed, people with periconceptual exposure to the famine.

Then in the bottom figure, you see the outcomes that we measured in this exposed group. I have laid it out so that the exposed group is right under the blue shaded area of the exposed group in the top graph. We looked at three outcomes. We looked at anomalies of the central nervous system at birth. These, in retrospect, are mostly neural tube defects and that is the black bar and you can see it peaks in the exposed group.

We looked at schizoid personality disorder, measured at age 18 in military recruits—all males—that came from this population. You see that also peaks in the same exposed cohort. Then finally we looked at schizophrenia in adulthood using the National Psychiatric Registries of Holland. That also peaks in the same exposed cohort. So, you have a very sharply defined exposure to famine, a periconceptual or early gestational exposure, which resulted in a marked peak in three neurodevelopmental outcomes at three different stages in the life course.

Whether those three outcomes have anything to do with each other, I can't tell you yet, except that they follow the same exposure. That was one study. It met a fairly high standard of evidence taken by itself in the sense that it was based on a historical event so people couldn't choose whether or not they would be exposed to the event of the famine. So, it is much stronger than your average observational study. Also, because it took place in Holland, where there was very good documentation of the food ration and the health of the population, and where decades later there were national psychiatric registries and other kinds of information that contributed to the strength of the design.

But you can never know from one study. There are always so many alternative explanations. There could be toxic exposures from other things that people eat when they were starving, like tulip bulbs, just to mention one. So, for a long time, we just couldn't be sure that this was a real finding. That has changed recently due to the work of another group. This is a group led by David St. Clair from Scotland and Lin He in Shanghai. They set out to test the schizophrenia finding of the Dutch famine study to see if they could replicate it in a completely different setting.

Their study was based on a massive famine that occurred in China in 1959 and 1960 after the initiation of the Great Leap Forward. They had much larger numbers than in the previous Dutch study, but they only had annual data, so in that sense the study was less precise. However, based on the previous Dutch study, they could specifically hypothesize that in years in which the birth rate dropped, the schizophrenia rate would go up, if periconceptual or early gestational exposure to famine was indeed linked to schizophrenia.

What you see here is in 1960 and 1961 the birth rate drops dramatically, and then you see there is a twofold increase in the risk of schizophrenia in the same birth years. It is relatively stable across the rest of the period. It is a very similar finding to the Dutch famine study and they were able to do this because they identified a region of China, the Wuhu Region, which is in Anhui Province, where, again, they were able to identify all the cases that occurred of schizophrenia in that area over a long period of time. So, it was like having a psychiatric registry over a long period of time.

Now we have a third study. Subsequent to the Wuhu study, David St. Clair and Lin He contacted me and we also brought in MaryClaire King, and the four of us are now working together to pursue this question in a joint effort. We have finished a third study in the Guangxi Region in China where we have the same result, but even stronger.

With three studies of this kind, I think we are fairly sure now that early prenatal famine is linked to schizophrenia in adulthood. But we still don't know why. We have a study with a fairly precise exposure, timing from the Dutch famine. Then we have the Chinese studies with very, very large numbers of cases. Together they provide fairly strong evidence, and the Chinese samples give us the ability to follow up these findings because there are very large numbers of people that we could now go and study and try to figure out what it is that explains this latent effect.

How do you go from something that happens in early embryonic development to something that may not happen until age 30 or 40? We don't know, but we are using biological reasoning. We are using available clues to guide ourselves here.

I am going to move now from facts to speculation, which I state up front to prevent any misunderstanding. I'll talk about some of these clues. There are several clues which suggest that the folate pathway could be important. One has to do with a gene which codes for an enzyme in the folic pathway, an enzyme called MTHFR. There is a variant in that enzyme, which has been associated with schizophrenia in very large samples. That leads us to consider whether there is something about the folate pathway that could be important.

Another line of evidence is that the risk of neural tube defects is known to be related to folate, that is, the risk is reduced by periconceptual folate supplements. The increase in neural tube defects was exactly coincident with the increase in schizophrenia in the Dutch study. There are also other reasons for us to look at the folate pathway. So, that is one of the places that we are looking.

How could the folate pathway be involved here? There are two hypotheses that I would like to mention to illustrate the way we can think about genes and environment together here. One is based on our knowledge that the folate pathway is very important in DNA synthesis and repair. Folate deficiency is thought to be one of the causes of *de novo* mutations. One way in which famine could be related to latent schizophrenia is that it is actually an environmental cause of genetic mutations. So, it is not exactly gene–environment interaction, but it is environment to gene to disease.

A second hypothesis is "epigenetic." We know that the folate pathway is also important in DNA methylation, which is one of the key mechanisms for epigenetic effects. We know from animal studies, for example, that maternal folate supplements influence the methylation of the DNA of offspring in utero. Since many people today have speculated that epigenetic effects are important in schizophrenia and there is some evidence along that line, this is another way that folate deficiency could affect the risk of schizophrenia.

You don't need to bother with the details of the diagram of the folate pathway there. I put that diagram there to point out these roles of the folate pathway for those who aren't familiar with it. If you look in the top left, you see pyrimidine and purine, which points out that folate is involved in DNA synthesis. If you look at the bottom, you see methyl transferases, DNA, RNA, which points out that it is also important in methylation of DNA.

So our general hypothesis is that folate deficiency might be a cause of *de novo* genetic or epigenetic events. It is just that, a hypothesis, but it is an interesting one. If it is true, it would have enormous implications for prevention and that is one of the reasons that we wanted to go after it. Here we can use the example of neural tube defects as the classic model, the sort of hope or the Holy Grail, because we do know that periconceptual supplements of folate do reduce the incidence of neural tube defects. We know that from randomized clinical trials. Our dream is we may be able to do that kind of thing for schizophrenia.

I don't want you to think that only prenatal famine or only prenatal experiences are important in schizophrenia, even when we are just thinking about the environment. I am pointing this out because I think in autism also one ought to keep a broad view, and the environment can have important effects at different times in the life course.

In schizophrenia we can begin thinking about the environment even before conception. We have good evidence now that people born to older fathers—in other words, the fathers are older at the time of conception—have a higher risk of schizophrenia than other people. The age at which men have children in a particular society is partly an environmental, sociocultural phenomenon. But we hypothesize here too that the mechanism linking this phenomenon to schizophrenia is *de novo* genetic mutation or epigenetic events. If the former is true, that could be considered as another example of an environmental factor leading to a genetic mutation. But in this case, the mutation could occur in the germ line even before conception.

After birth, we have evidence suggesting that social factors influence the risk of schizophrenia. There is strong evidence nowadays that certain immigrant groups in western Europe have very high rates of schizophrenia. Urban living has an effect on the risk of schizophrenia. Very probably, so does cannabis use. These things are not mutually exclusive. We probably will find that it is not only one point in the life course that is important for the cause of these disorders.

What can we say that might be useful in terms of autism? David asked us each to draw from our experience with these other diseases and say what would we suggest could be applicable to finding the causes of autism. There are three things that I would take from the experience with schizophrenia. One is I think that we should create and support autism registries. This is already being done in some places. There is a good registry in western Australia and I know people are talking about doing this in the Scandinavian countries and to some extent in the United States, but I think that is going to be key.

Second, I think that we need to establish what I call pregnancy birth cohorts. These are very large populations of people on whom we have measured their prenatal exposures. So, we have archived prenatal and cord blood samples, for example, that can be used over a long period of time to measure environmental toxins, nutritional states, and infections, and we also have genetic information on these people and their mothers and fathers.

You have to actually begin early in pregnancy in order to collect the information that you want to collect. You can't do it retrospectively. You have to do it prospectively and you need very large numbers. It is possible to do this. It is already being done in one large cohort of 100,000 in Norway. Within that cohort we have the Autism Birth Cohort or the ABC. Maybe Ian Lipkin and Allen Wilcox will talk more about this later. Then we are also starting, in the United States, the National Children's Study, as Phil Landrigan mentioned. So, there are two examples where I think we have studies that are going to yield some answers to these questions for autism.

Finally, along the lines of the example that I showed and the comment that I made earlier, I think we should look for—we call them natural experiments. I am not sure if that is a misnomer. But we should look for historical events that result in people being exposed to harmful or protective factors and we should go to those places and study those people. There are so many opportunities to do that once you recognize the design as a useful approach to study these diseases.

Dr. Schwartz: Thank you very much, Dr. Susser. Are there any questions—I guess we can take just one question for Dr. Susser.

Dr. Beaudet.

Dr. Beaudet: Can I ask if there is any hint that the incidence of schizophrenia might be dropping as we—looking at the parallel incidence of neural tube defect, but with a much later age of diagnosis?

Dr. Susser: Well, it is a controversial subject. Some people think that the incidence of schizophrenia is dropping and others think it isn't.

We don't actually have an answer to that, but the time to look would be a little bit later, you know, because it is 20 or 30 years after we started with the fortification of folate and so forth. So, we need to wait another 10 years to really get that answer even.

The other thing that I would say about that is that somebody should follow up the randomized trials that were done of folate supplementation to prevent NCDs and look both for adverse effects that we may not have known about and the good effects like reduction in schizophrenia in those studies. We do have these randomized trials and the people are already in their teens now, who are in them.

Dr. Beaudet: I will show some data just to suggest the increase in folate was going on in the seventies and eighties and didn't only occur on into the nineties and so on.

Dr. Susser: Yes, that is true, but the fourth vacation was introduced then.

Dr. Wilcox: Ezra, in the context of natural experiment, it is interesting that in Norway and maybe other Scandinavian countries, they have relatively low natural levels of folate in the diet and a great resistance to taking something artificial like vitamins. So, there are national differences that could be exploited for this kind of question.

Dr. Susser: Exactly right. We intend to do it. I know you have done it to some extent in very effective ways.

Dr. Schwartz: Well, thanks again.

Our next speaker is Fernando Martinez. Dr. Martinez is a true environmental scientist. He is a pulmonologist, a pediatrician, an epidemiologist, and a geneticist. He is an integrated investigator all unto himself.

He is the director of the Arizona Respiratory Center at the University of Arizona as professor of pediatrics there, and he is going to discuss some lessons learned about environmental asthma.

ASTHMA[9]

Dr. Fernando Martinez

Dr. Martinez: I hope that the examples that I will produce today are

---

[9]Throughout Dr. Martinez's presentation, he may refer to slides that can be found online at http://www.iom.edu/?id=42463.

kind of intermediate between what Dr. Susser has shown and Dr. Landrigan has shown in terms of our degree of understanding. I would like to say that perhaps they have to do with two issues that have been raised, natural experiments and the potential role and interests that protective effects may have. So, what I am going to talk about is what I am interested in, which is asthma. You are going to see things here that may sound very familiar to you. Asthma is a heterogeneous set of related conditions in which recurrent, partially reversible airway obstruction is the final common pathway.

The clinical expression of asthma can start at any age, but we have now found in the last 5, 10, 20 years, that the first manifestation of the disease usually occurs during the preschool years. That may sound also familiar. In our case we have well-defined intermediate phenotypes for asthma that are strongly related with the disease burden and therefore they can be studied separately and I will show you some examples of that. For example, aeroallergy, bronchial responsiveness, or total serum IgE. One thing we know about asthma is that in the last 40 years it has clearly increased in frequency and you can figure that out both through the diagnosis of asthma and through asthma symptoms as reported by parents. So, there is a strong hint that asthma is an environmental disease and, of course, we have an advantage, I think, with respect to autism in that asthma is a very variable disease, an extremely variable disease. We have known for years what the main triggers for the disease are, and, of course, some of them have been pursued as potential inceptors of the disease. In other words this is a concept that is very important. In asthma we know very well that there is a difference between what could cause the disease at its very beginning and what triggers the disease once the disease process has developed.

Unfortunately, we have not been very able to show for any of the triggers that they are involved in inception of the disease. The one that I am involved with and that is the only reason why it doesn't have a question mark—it should have also a question mark, but we are all biased, of course—is the lack of certain protective effects. That is the one I am going to stress more today.

It all started with a form of natural experiment. A researcher in Britain, David Strachan, working with one of the largest birth cohort studies as two of the previous speakers have talked about, the 1958 birth cohort in Britain, found a startling finding, which is that children who had older siblings at home were much less likely to have what could be considered intermediate phenotype for asthma, which is hay fever, than those who did not.

That was truer for those who had older siblings than for those who had younger siblings at home. This observation was ignored I think for years until we, in our own longitudinal study, which was started in the 1980s, tried to reproduce it. This, as you can see, is 11 years later. What we found was very interesting. Here are "in the triangles" the children who were exposed to other children, be it because they had older siblings or because they were taken to day care.

As you can see here at the beginning of life, they tend to have, of course, more viral infections, which are strongly associated with wheezing with this age period. But very interestingly if you follow them enough by the sixth to seventh year, when the atopic form of the disease, the allergic form, starts to be more prevalent, these children are clearly and significantly protected.

A series of other studies came showing that there were other protective exposures. For example, this has now been reproduced, replicated in 10 studies and not replicated in 2. If you have a dog in the home, you are less likely to develop asthma in the first years of life than if you do not have a dog.

Now, what is common between having a dog in the home and being exposed to other children? Well, in several studies now, it has been shown that day care and homes with a lot of children have high concentrations of a marker of microbial exposure in the homes, which is endotoxin. This has been shown now repeatedly in many studies. This is true for pets, as it is true for day care and homes with heavy concentration of children.

But perhaps the most interesting solid natural experiment is the one that you see here in this slide, which is a form of living, which still exists in Central Europe in which children and adults live in single-family farms as the ones you see here.

In these single-family farms, an empirical observation was that there was really very little asthma. Researchers listened to local physicians who were telling them that there was very little asthma in this environment and went to study it. What they found only a year after we published that paper on day care and other siblings was that, lo and behold, both for subjective and for objective measures of asthma, the children who live in those farms that you saw there were between 5 and 10 times less likely to have asthma than those living in the same rural communities, but away from those farms.

They also studied endotoxin concentrations in the homes, which is here in the x axis in relation to the likelihood of having illnesses in this environment. As you can see here, immediately after the first publication about those farming environments, what they found was that there was a striking inverse relationship between endotoxin exposure and the likelihood of having asthma, particularly allergic asthma, but not nonallergic asthma, which is an important issue because the clinical expression is identical. It is impossible to distinguish them.

Even more and I don't show this slide to take more time, but in nonatopic asthma the relation is inverse to this one. In other words the more endotoxin, the more nonatopic asthma. We will get back to this concept in a moment. Of course, this was a very extraordinary environment. So, it was necessary to try to reproduce this in a less extraordinary environment. If you want to consider Manchester in England a less extraordinary environment, here it is.

As you can see, these researchers in Manchester clearly reproduced in an urban setting the findings that had been reported before for the extraordinary environment in rural communities in Europe, in Central Europe. Now, an explanation has been proposed for this association, which is very simply as you can see here, that endotoxin or LPS increases the expression of IL-12 and IL-18, which in turn has a downregulating effect on Th2 differentiation, T helper cell 2 differentiation, which is the central and most important determinant of having atopic diseases. So, it was proposed that if you have endotoxin exposure, you deviate your immune responsiveness away from the Th2 mediated response, which is responsible for atopic asthma. If you don't, as you see on the right side, you upregulate the likelihood of having an atopic response.

Asthma is also an allergic disease, and here I have put the latest twin studies published by the same research group with respect to asthma and interestingly with respect to autism, just published this year. You can see there are many things in common between the genetics of asthma and autism. I think the reason why the validity of asthma appears to be greater is simply because this was done earlier in life and the twin studies of autism were done later in life or perhaps because it is true since I know very little about autism, I don't know the answer.

Something very interesting and paradoxical, however, is that these twin studies have both shown no shared environmental influences affecting the concordance of asthma between twins, which may be true, but may also be a complete artifact, due to the fact that the models used suppose that there are no gene–environment interactions. I will get back to that concept in a moment.

But much like in the disease of your interest, in asthma what we have had is that no single study has shown strong statistical evidence of a single gene being responsible for the disease. We have 15 chromosomal regions in which there seem to be asthma genes. Sound familiar? And only three regions have been clearly reproduced in at least two studies, if not three. Same thing as for autism.

Well, one of the reasons why people have been able to reproduce linkage with asthma in chromosome 5q is because there is, I think, a large array of potential genes that can be candidates. It is a problem of luck. One of those genes is CD14, which I showed before. CD14 was in the middle of this potential pathophysiologic explanation. Why? Because CD14 is one element that is a member of the receptor system for LPS.

CD14 is a crucial member of the receptor system for the exposure and protection in the exceptional farming environments and also in the nonexceptional environments in Manchester. We sequenced that gene in populations and we found five main closely linked, single-nucleotide polymorphisms in the five prime regulatory regions, and for one of them we showed functionality. This morning, we were told that that was important and I am just showing one slide of probably 10 I could show about the functionality. You will have to believe me that transcription rates are increased in carriers of the T allele at position –159. We now know that this is a problem of balance in those particular 5' regions between SP1 and SP3 transcription factors. It was logical to suppose that if you had more CD14, you would be more sensitive to the environment and you would have less atopy. There was more atopy in our population among children who were CC and CT, who had low expression of CD14 and those who had TT. We thought we had to put ourselves in the hall of fame of geneticists, who had found something important until, of course, we fell into the same problem that every single other person working with complex diseases has fallen into, which is that three researchers were able to reproduce this and three researchers were unable to reproduce this result.

The three researchers who reproduced us called us and congratulated us. The three researchers who were unable to reproduce us said you guys don't know what you are talking about. Of course, we immediately thought that the right place to study this was where the exposure to endotoxin was the highest, the farming environments, because here perhaps if we could determine if these people who live this way and heavily expose their children to brown stuff that is here—about which I won't talk before lunch—could probably heavily expose the children to endotoxin and others could not be exposed. So, we could study gene–environment interactions. Of course here just to show to you how these things work is the relation between the same polymorphism and atopy without considering the environment among farmers, there is nothing there with CD14. The trick was to put it in relation to the environment and when we did that, something very interesting happened, which is that the sensitivity to the exposure to endotoxin was completely different, depending on your genotype.

The CCs were heavily sensitive to the environment. The CTs and TTs were not. Now we have shown in other functional experiments that baseline unstimulated production of CD14 is higher in TTs, but a stimulated production is higher in CCs and CTs. So, what happens is that you have a very flat line for TTs and CTs and a very steep line for CCs. That creates a very interesting paradoxical situation, which is that at lower levels of exposure, the CCs are at risk whereas at high levels of exposure the TTs are at risk and the CCs are protected. This is due to the fact that the genes don't act alone.

If you don't believe that this is true—and I would agree if you see only one study, which was done in the exceptional environment—this next slide shows the results of the same analysis done in Manchester and the result is exactly the same. The CCs show this very steep relationship between risk of being allergic and the exposure to endotoxin. TTs and CTs show much less response to the exposure to the point that CC risk is lower at high level of exposure. Among African American adults, the same thing has been shown by Williams and co-workers in Detroit.

What are my proposed conclusions? From our experience, natural experiments are very important and they may be true for both risk exposures and protective exposures. In our case it was protective exposures and they provided significant cues or clues for us to understand. I think I would have to say the hygiene hypothesis as this is called is still very controversial, but I think it has focused us into an area of exposures and has allowed us to understand the disease much better. Not only that, it has inspired new treatment. Very recently in the *New*

England Journal of Medicine, a paper has been published in which ligands of TLR9 are used as adjuvants for allergic desensitization, with the idea that they wanted to reproduce a little bit what could be present in the environment in this particular condition.

I believe that we have to understand biological systems as plastic with heterogeneous responses to the environment and I think the example of CD14 that I have presented to you is characteristic of what the complex related genetics are going to be. They are going to be nonlinear. They are going to be weakly linked and strongly context dependent.

Suggested approaches: I think that following up on replicated enhancing or protective exposures may prove extremely rewarding. I am not an expert in the field, the very controversial but very interesting fact that Mexico-born mothers have children with less autism than those who are not, maybe because I am Hispanic, too, calls very much my attention. I do understand that this may well be due to bias because they may seek less access. They may recognize this less, but being a physician who works with a lot of Mexico-born mothers, I am quite aware of how worried they are about the health of their children.

So, I am not very convinced about their argument. I think that we have now technologies, both at the genetic and epigenetic level that allow us to assess genomewide the potential for genetic and epigenetic factors to be present. That may be related to exposure. So, I think that in studies such as the National Children's Study, we could determine if replicated exposures could be useful to determine the type of gene–environment interactions that, only in an example, I have provided to you in the case of asthma.

Thank you.

Dr. Schwartz: Fernando, that was great.

Let me just make a suggestion, then we will open it for group discussion and focus on the general topic.

DISCUSSION

Dr. Schwartz: Your talk, Fernando, brought up a really important point, which is that environment can be used to narrow the pathophysiologic phenotype in such a way that you can understand the genetics of and also potentially the biology that underlies a very complex disorder like asthma and consequently a complex disorder like autism.

But I wanted to ask the group in general about natural experiments and whether we can expand our concept of natural experiments to autism. What are the natural populations or the cohorts that might be available and amenable to further study? Can we follow as it relates to an autism endpoint or subclinical early condition that is along the pathogenic line or clinical line or development of autism?

Phil Landrigan, you brought up and, Ezra Susser, you brought up the issue of the natural experiments. Are there populations? Clearly, the National Children's Study is going to be an outstanding study that will allow us to follow kids over time through development, but it is also going to take a long time. Are there populations that have been exposed that we should be looking at more carefully for autism influence?

Dr. Landrigan: David, a couple of responses to that. First of all, I think the National Children's Study is very powerful, but it is probably not going to answer every question. When I think of natural experiments, maybe because I spent many years at CDC, I think of clusters. I think of Brick Township in New Jersey, for example; I think of the group of children now three or four decades, who were exposed in utero to thalidomide and I think there is need for highly focused studies, which look at children who suffered unique exposures. I also think it is terribly important when those kinds of studies are done that we do as Ezra Susser suggested in regard to big cohort studies. That is, that we take samples and we archive them because there is always the very high possibility that new diagnostic techniques or new genetic probes will be developed in future years that will enable the scientists that follow us to examine those specimens and ask questions that are not possible to ask today.

I think with regard to the Children's Study, I would say that it won't be that long. It will certainly be in our lifetime that we have data on the relationship between the environment and autism now that federal funding has been made available by the Congress and in such a way that it doesn't destroy the budget of NICHD. We are going to be moving forward. The first recruitment will take place beginning in about 12 months. That means that we will have a large number of 3-year-old children in the study in about 5 or 6 years, something like that, 7 years at the most.

So, I would argue we will begin to make data on gene–environment determinants of autism in that particular population available by 2010 or 2011, 2012, somewhere in that range, not tomorrow, but not 25 years either.

Dr. Lipkin: I am really not going to be talking much about the Norwegian cohort because I was asked to talk about infectious diseases. But, in fact, that study is well underway and it has been running now for several years and at present we are close to 80,000 children recruited. I would think that the time frame for having real data there is much shorter. It is in the neighborhood of 1 to 2 years. That study includes prenatal data on the child, on the mother. It includes genetic information on the father, on the mother, on the child, cord blood, urine, a wide variety of sample types.

One of the problems and challenges that we face right now is that although there are resources that have been allocated for establishing the cohort and for collecting these samples, that funding will be expiring in the not-too-distant future. Furthermore, there is really no allocation as yet to do any sort of work to analyze environmental exposures or to look for biomarkers, anything of the like.

Now, there are a number of people who are here who are working with that cohort. Ezra is involved with this. Alan Wilcox, Mady Hornig, and myself. We would encourage people to collaborate and begin using this resource as soon as possible.

Dr. Schwartz: Is this a population that is large enough to look at autism as an endpoint?

Dr. Lipkin: It is 100,000 children.

Dr. Schendel: There is also a Danish cohort that was assembled beginning in the mid-1990s of 100,000 pregnant women followed up, including their children. That is a database that is available. It doesn't have the intensive clinical evaluations that I think are being funded for the Norwegian cohort, but it does have biologic samples collected at multiple time points in pregnancy and cord blood of the children and newborn blood spots and has baseline data of the features of the mother during pregnancy and postnatal development of the child, which is clearly a resource that could be used in tooling our *de novo* studies.

Dr. Lipkin: Just to make one point of distinction, the Norwegian cohort, which followed the Danish cohort, actually has materials that have been collected specifically for proteomic analysis and transcript profiling. It was really connected primarily not to do only genetic studies but to really look at functional data, so that rather than having blood spots, we actually have materials that are stored at minus 70 degrees and really, I mean, the opportunity to do proteomics is really going to be unparalleled until such time as the National Child Study comes online.

Dr. Schwartz: So, it sounds like a terrific population. What are the limitations? Are there limitations within this population?

Dr. Lipkin: Of course, one issue is whether this population is too isolated to tell us what is happening in the American population. That is one issue that has been raised.

I don't know. Would anybody else like to speak to this?

Dr. Schendel: I would like to throw out to the group a question that I wanted to raise in the first session this morning. It might also apply to Session II and certainly to Session III as another opportunity for comparisons which might serve as a natural experiment, which is the sex bias in autism, the fact that you have this extensive male bias, but the extent of the male bias varies depending on the phenotypic profile of the group, with girls obviously displaying autism much less frequently than boys.

I am throwing this out to the speakers of these sessions. Is that an opportunity that we can use for an investigation of clues for protective or risk mechanisms for autism?

Dr. Insel: One example of that which shows up in the recent literature is a point that Dr. Susser made about paternal age as a factor; much greater odds ratios when you look at girls with autism than boys, about 18-fold increase that your father will be over 40. In girls about a fivefold increase with boys.

Dr. Schendel: My point is using that dichotomy between boys and girls as a field for identifying potential mechanisms might explain the susceptibility to autism.

Dr. Slotkin: I actually think that is not going to be fruitful because basically that sex difference exists for almost all neurodevelopmental disorders and it is likely a reflection that you can have the same degree of initial impact, but the female brain is more plastic because estrogen receptors regulate plasticity. So, you guys have an advantage over us apparently. I don't think that is something that is going to be a fruitful etiological factor for autism because it is simply shared by everything from ADHD to physical trauma of the brain.

Dr. Beaudet: I would like to just strongly disagree with that opinion. I think that there is a lot of evidence that genetic aspects of females and males with autism are very different and I will show a little bit of data and some speculation about this, but I think that the causes of autism in females and the causes of autism in males are very different, I believe.

Dr. Schwartz: Art Beaudet, can you expand on that in terms of environment?

Dr. Beaudet: It might be better to wait until my presentation when I will show a slide which would be much better to address the question.

I think the boys are much more likely to have major environmental factors.

Dr. Alexander: We are really in a very fortunate position, I think, to have the potential for three large national studies for environmental influences on children's health and development from different environments. The Scandinavian environment probably has significant differences from what we experience in the United States. That will be an interesting comparison in and of itself.

Autism is clearly one of the major outcomes that we are going to be looking at in the National Children's Study and as Phil Landrigan pointed out, it will take us about 4 years to recruit the hundred thousand sample as those kids age—by age 3 we will have basically all the autism kids diagnosed and identified.

So, we will be able to start looking at the analyses we intend to do with regard to autism for the whole cohort within 7 years. One thing that relates to some of the discussions this morning, where it was pointed out that many of the either biomarkers or the toxicants are looked at singly and at one point in time. The advantage we have here is we will be able to do comparisons—analyses of interactive effects potentially with multiple exposures at several different points in time and also in relationship to genotype. That is an additional factor that we are going to have going for us in the National Children's Study that I think is going to help shed a lot of information on the questions we are asking today.

Dr. Akil: This is probably silly and I am going to sound like a hippie, but that is okay. Old hippie.

I was struck by the discussion today about how in a way we are getting an evolutionary message. It is an evolutionary experiment in that the environmental factors that we can handle well, like living with a cow, maybe because we evolved in some kind of selection so that we could cohabit with a cow. If anything, when it is protective, it has an advantage; whereas insecticides that somebody synthesizes we have no evolutionary advantage in protecting ourselves against them and they seem to be quite hurtful.

I am thinking about these gene–environment interactions and how to group things, it might be helpful to kind of have this very general idea about whether it is something that humanity has coped with in the past versus if it is newly introduced.

So, I apologize for the very nonscientific take on all of this.

Dr. Susser: I didn't want to leave your question hanging about opportunities for, quote, natural experiments and just mention a few examples. One I think that would be important would be to follow up on populations exposed prenatally or in early life to infectious diseases, which may be outside the United States. There were early findings that related rubella infection in utero to autism; no one knows if it is a true finding or not. We don't really have prenatal rubella in large numbers in the United States now, but you have massive epidemics in other countries.

Another example would be populations exposed to toxins in industrial disasters, also common outside the United States, maybe in the United States. There are many examples one can find and the only other thing I wanted to say on that is that we should also look for positive things that happen to populations, and use those as experiments to see if they have a positive impact on autism.

Dr. Schwartz: Let me ask very specifically, is the Agricultural Health Study a population that we could use to look at this carefully in? This study is a rather large population of, I believe, 60,000 to 100,000. Alan Wilcox, could you tell us something about this?

Dr. Wilcox: This is a cohort that was set up at our institute of agricultural health workers, mostly men, but some women, men, and their spouses, farmers and their spouses, who apply pesticides and I think somebody may know better than me, but it is about 60,000 people—90 were enrolled. Well, okay.

So, the family members have been also enrolled, but in smaller numbers. So, I guess the question is whether anything is being done with following the neurodevelopment of those kids. You were on the advisory panel for that, weren't you. Do you know?

Ms. Bono: I was on the panel, the advisory panel for about 6 years, but I haven't been for awhile. My understanding was that the number of children and wives was actually not as large and I don't know how much information. I know neurodevelopmental disorders at one point was on the panel of things we tried to look at. But I am not sure where that went.

Dr. Schwartz: The question is could we expand it to that? It would take an investment, but is it worth the investment to expand that cohort so that we could find out whether pesticides in agricultural chemicals are important in the development of autism.

Ms. Singer: I have a question for Dr. Landrigan.

Acknowledging that there is still work to be done with regard to lead, I think many of us would look at lead as a real success story. I was hoping that those of us who are parents, as Lynn said, feel a real sense of urgency. I was hoping that you could share with us some of the factors that will speed up your results and also that slow down your results to the extent that they apply to autism.

Dr. Landrigan: First, I would caution you not to proclaim that lead poisoning is over. There are still many tens of thousands of kids in this country with elevated lead levels. Yes, we have knocked down the average by 90 percent, but there are still pockets of kids, principally minority kids living in inner city old housing, who have terrible exposures, not just immigrant families. So, just to clear the record on that.

With regard to the factors, I think the biggest factor that sped up the discovery of the developmental neurotoxicity of the organophosphate pesticides was the decision that a number of the federal agencies made beginning in the mid- to late 1990s to substantially increase the investment in studying the impact of chemical toxins on children's health, with a particular focus on brain development.

I think in a lot of ways that is the take-home lesson I would like to give you from my little talk, that the lead studies, which languished for decades, were terribly slow to produce results and by contrast the pesticide studies, which were really very generously supported and spanned the gamut from the most basic science through clinical research to epidemiologic studies all the way through to intervention studies, yielded some dramatic results in less than a decade.

Dr. Martinez: If I may add to that, given our own experience in asthma, researchers as a collective tend to be quite conservative with respect to knowledge. It is difficult to put into the collective brain of the research community completely new ideas. There is a difficulty with funding—using the very limited funds that exist for things that are of very high risk.

For that reason, I think that the new approach that the NIH has taken to fund more risky research is extremely important. It is also a problem of our way of thinking as scientists and you have to justify that a little bit because there are a hundred ideas—I say a hundred, could be a thousand new ideas—that come up in a single year and you know that two or three are going to be successful. You have to be very lucky to be working with the one that is the successful one. I could have showed you many examples of unsuccessful ones in our field. So, it is just a problem of trying collectively to find those that are most promising and being willing to fund risky research that is at least in part solidly based on the knowledge we have today and at least replicate it. That is the other thing, not just one of the—or two, but replicate it.

Participant: I am Wendy Harnisher. I am the parent of two boys with autism. I was in the overflow room, so I actually want to bring up something that I heard before the break.

I liked something that Susan Swedo said about Tylenol and vaccines. It is my understanding that Tylenol reduces glutathione levels in our children and glutathione is responsible for pulling toxins, including heavy metals, out of the body. I think this is something we should seriously look at.

Also, I wanted to comment on something that Dr. Pessah said about the calcium channel being a main target for mercury. I think we need to look at mercury as well as other heavy metals. I am currently chelating both of my boys and they are dumping a lot of lead and they are also dumping a lot of mercury and with each dump, they are getting better.

I think we should look at populations that do not have autism, such as the Amish, but not just the Amish. There is a pediatrician in Chicago who claims that none of his patients have autism. I know we are not supposed to bring up vaccines here today, but he doesn't vaccinate his kids.

Dr. Leshner: Thank you.

Dr. Schwartz: Thanks for your comments. Really appreciate them. I do have a question for the group at large because Phil Landrigan brought up this issue of how subclinical early indicators of response to lead were helpful in identifying safe levels of lead and also moving the research forward. It made me wonder whether there are subclinical phenotypes or subclinical, preclinical biological responses in autism that would help us identify etiologic agents.

It is a full-blown disease or there must be spectra of this disease and I guess the question is should we be looking at any of those less severe forms of the disease?

Dr. Levitt: In the context of the several studies that have just come out on the examination of baby siblings, there are several studies that have come out where they have looked at what you might call intermediate phenotypes or whether there are clear indications that there are atypical trajectory of social, behavioral, development, and communication. I think the answer from several studies is, yes, there are and you can identify those; that is, skilled individuals doing research in that area can identify that there is a typical development and trajectory in those domains. Now, how that relates to whether those children are going to end up on the spectrum or not, I think is an open question, but I think—when were the baby sibling initiatives started historically? How long has it been?

Participant: About 5 years.

Dr. Levitt: So, 5 years. So, the first studies—the Davis study, the study that came out of Vanderbilt, there is a third study. All three basically show very similar things with reasonable numbers so the statistics I think are okay—indicate that you can actually begin to think about doing that where you could identify individuals, children, young children before the full-blown diagnosis that then would need to be followed, but it has been 5 years and now we are just getting the first indications that this might be a fruitful way to go.

Dr. Insel: David, I heard a slightly different message in the talks and if we could go back to Dr. Martinez's comment about the two forms of asthma and how if I heard you right, you said they were clinically indistinguishable and yet you had a very different pathophysiology pathway. How did that happen? How did you get there, knowing that you couldn't do it just from the clinical phenomena?

Dr. Martinez: What helped most is what you could call biomarkers in a very generic way. In other words we learned. It was tough because there was a period during which the lumpers had the prevalence or the splitters among the scientists. So, everything is asthma. Treat it with inhaled steroids and everybody is going to be okay. But with time we started learning that if you take, for example, responsiveness or you take being sensitized to—or not, those are very different kids or adults, who have the same symptoms. Of course, if you go into a lot of details once you know they are different, you start seeing that there may be differences, but it is very difficult to do that before you have the specific biomarker that allows you to do that.

So, in a certain sense I was hearing before with respect to efforts of phenotyping, I think the efforts of phenotyping, somebody said that before—I don't remember who—include the biomarkers. It is not that you can start by trying to squeeze your brain to distinguish clinical characteristics and that is the only way you are going to do it. You have to include the biomarkers.

We found out, for example, in these studies that children who are skin dispositive to allergies are very different from those who are not. They have a different prognosis. They are much less chronic. They tend to decrease with puberty. I won't go into all the details. But we knew that once we developed the biomarkers to understand how to distinguish the different groups and that is something that could help a lot in this particular area, too.

Dr. Schwartz: And the environment helped distinguish the different groups. So, there were several factors, biomarkers, genetics, and environment.

Dr. Martinez: There is no doubt. It is an integrated process. It cannot be done one first and then the next.

Dr. Hertz-Picciotto: I guess I wanted to sort of link some of the discussions that we had earlier with the discussions just now and raise the question, so, we have these large birth cohorts, which have a lot of promise in terms of being able to go back because there will be stored specimens that we can do a lot of varieties of biomarker testing on. On the other hand, the question of the actual characterization of the phenotypes in those very large cohort studies, I think, is a concern that needs to be raised because the diagnosis—well, I don't know very much about the Norway situation and what is being done in terms of the phenotyping and just even in terms of the ascertainment of the diagnosis itself, but it does seem clear that from our study, from the CHARGE study, that there is a substantial percentage of children who don't meet criteria for the condition. Maybe those are the ones that fall into this category of having some of the markers and are still informative, but I think it is important that even in terms of knowing who actually has autism and who doesn't in these large studies, it is going to be a big challenge in the National Children's Study having just spent the last 6 weeks writing one of those proposals and working with some of the counties that don't have academic centers and I suspect don't have a whole lot of people doing—with the expertise to, in fact, do the diagnosis in those areas. So, I think that is another challenge and in the National Children's Study, one of the things that I noticed was not part of the RFP and not part of the protocol is anything prior to 36 months in relation to autism. So, there is no screener that is happening at 12 months, 18 months, 24 months that is part of the current protocol. It is an area where I think some work is going to need to be done.

Maybe that will identify a large percentage of the ones who at 36 months will meet criteria and maybe it will identify some of the subclinical conditions that may or may not go on to, in fact, be autism after the 6 months and could be done on the 100,000 children, conceivably. Not to mention the subphenotyping issues that were brought up earlier in the first session this morning.

Dr. Herbert: There are some papers by Dr. Deborah Fein's group at the University of Connecticut on children losing their diagnoses and it is informative potentially to look at what they lose—what they have left after they lose their diagnosis. One group had specific language impairment, attention deficit. Does the way that the phenotype decomposes in the course of treatment tell you something about how it is stuck together?

There has been work on the idea that the different behavioral traits transmit separately. But we really don't know what that means biologically. Is that purely a matter of genetics? Is it a matter of other biological issues? Is it a matter of gut bugs, a variety of things? In any case, I think if we had biological measurements in the course of tracking the progress of people in treatment, we could learn something.

I want to make one last comment, which is that in order to gather this data, we are going to have to have some tolerance for exploratory measures, where we aren't exactly sure what it is going to show, but that this is a good time in history to take that on.

Dr. Schwartz: Great comment.

Sallie.

Ms. Bernard: Couple of things quick. One is just to your point. It might be interesting to look and see how autism composes in addition to decomposes because with my son, he first had language and then attention deficit and then autism. So, PDD and then autism. So, that is sort of an interesting rule-out.

Also, I want to go back to your question about what populations we can study and you know, while these big cohort studies, the Norwegian study and the National Children's Study, are vitally important and we need to do it, we do need to keep the idea of urgency and there are populations that exist right now that we could be studying that focus on autism. The speaker from the other room brought up the Amish. We have talked about the baby sib studies and those are very specific populations that we could go in very quickly and study the rate of autism and look at exposure histories in those groups and see—get some good information right away without waiting 3 to 4 years and spending a huge amount of money.

The last thing I would like to point out is that we focus very much on complex diseases with the idea that it is a foregone conclusion that autism really is a complex disease and there are multiple genes, 10 or 20 genes and there are a thousand exposures that could be possible. I would just like to remind us that historically there is a disease called acrodynia or Pink disease that was one of the number one childhood diseases in Australia about 50 to 75 years ago. It had one cause, and it was mercury.

I just want us to think about and not rule out the possibility that the causes of autism could be more limited to what I have heard in the discussion today.

Dr. Schwartz: Thanks very much. One last comment.

Participant: Mary McKenna, University of Maryland School of Medicine.

I am a neurochemist and I was very struck by Martha's talk earlier and also by listening to the other people talking. Many of the nutrient and alterations in metabolism that have been brought up, for instance, folate, B_{12}, B_6 are very important for brain development and it would seem to me that it would be extremely useful to come up with some sort of metabolic panel for testing for any kid suspected of autism, where you would look at all these vitamins that are possibly at risk, where you would look for heavy metals and you start screaming right when any sort of diagnosis is first suspected because you may come up with a lot of useful information if you start obtaining things like that.

It doesn't seem like there is any sort of uniform consensus right now as to what the approach is and what biomarkers and what metabolic information and heavy metal information to get right from the beginning.

Dr. Schwartz: One last comment, Laura. Make it short, though.

Ms. Bono: I just want to mention something about what may be called the "hit and run" and that as we are gathering urine and blood specimens on children, it is very hard to perhaps track the toxicant that hurt the child. Speaking to metals, which is the same general idea, before we started chelating my son in 2000, we ran a test at Duke and we got a 5 on his blood lead levels. But he has dumped more lead and mercury and aluminum and nickel and tin, which, of course, points to the synergistic effect of metals more than any kid I have ever seen. I mean, lead levels off the chart every single time and he started out with a blood level of 5, which leads to the theory currently of some doctors that these kids are non-excretors.

So, you wouldn't find that in the blood when you first start testing them. The blood is not the organ of toxicity. The mercury and lead and other things are going to other areas.

Dr. Schwartz: Thank you.

Dr. Leshner: Thank you, Dr. Schwartz and the speakers and all the participants. I think we have had a tremendous morning. I have had a tremendous morning. You can speak for yourselves. I think it has been very interesting. I like very much the spirit of leaving all the questions on the table; that is to say to take as broad a conceivable look at this as we can because as we go forward, although by necessity individual scientists and groups will develop their own priorities and their own specific projects. I think there is no question that as a field or at least as an outsider to the field, listening to it, sure does need to leave the field as open as possible and there is a tremendous amount of research yet to be done and to be discussed.

Session III
Environment and Biology I: What Are the Tools for Autism—What Do We Have, What Do We Need?

Dr. Leshner: As far as I can tell, there is general agreement that it has been a terrific morning, and therefore we are putting terrific pressure on the afternoon speakers, so don't let us down. People have been very well behaved and stayed on target, asked questions, didn't make long speeches, so everybody so far has behaved very well, identified gaps.

Dr. Levitt: Good afternoon. My name is Pat Levitt. In this session, there is one content session and then the discussion; this is the session that has the unenviable task of putting together environment and biology. That is the title of it.

The first speaker is Art Beaudet, who is professor and chair in the Department of Molecular and Human Genetics at the Baylor College of Medicine.

HOW MAY ENVIRONMENTAL FACTORS IMPACT POTENTIAL MOLECULAR AND EPIGENETIC MECHANISMS?[10]

Dr. Arthur Beaudet

Dr. Beaudet: I am going to maybe be a bit provocative and try to argue that there is a substantial chunk of autism where we now can predict what is going on. I made a diagram here. You heard earlier this morning the mention of maybe 10 percent of autism being genetic.

I'd like to argue that this is 40 or 50 percent. This is maybe an exaggeration, maybe it won't be quite that high, but there is quite a group that we know where we are going to end up.

These individuals have chromosomal defects and single gene defects. You have seen some of these mentioned. You have heard someone mention how we have better and better techniques for how to

[10]Throughout Dr. Beaudet's presentation, he may refer to slides that can be found online at http://www.iom.edu/?id=42981.

search for these. But we know the way we search for them today is like looking at icebergs and only looking above the water line. So we know that this group is much larger because of our inadequate ability to detect small changes.

Then I would argue there is a second group over here that is much more unknown as far as what is going on, and much more likely to be a candidate for involvement with environmental interactions, epigenetics, you name it. I think we just don't know where we are there.

This is a review from about a year ago that is very nice, indicating how many known definitive chromosomal abnormalities are seen in autism. These are mostly so-called *de novo* events in children. Their parents are normal. Duplications of chromosome 15 are by far the greatest, and there are quite a lot of deletions of chromosome 22 as well. So we know there are deletions and duplications that can involve every single chromosome that can give rise to autism.

There are a couple of papers that have appeared recently that further emphasize this, using a methodology called array comparative genomic hybridization to detect larger events across the genome with greater efficiency. One paper recently reported detecting abnormalities that are presumed causative in children, with 27 and a half of the children with syndromic autism. These are children who are dysmorphic. They are likely more cognitively impaired, probably both mentally retarded and autistic in most cases, and unusual looking. If you see them in a grocery store, you will see that they have some physical abnormalities.

The report by Jonathan Sebat has been mentioned, where he found in about 10 percent of simplex cases these kinds of abnormalities. We know that these methods being used will miss many, many kinds of genetic lesions which would give the same functional effect.

We have heard about advanced paternal age. Dr. Susser has this publication here. I will just say that we have an ability to make a very good guess what the problem is with advanced paternal age. It is probably point mutations, so it is probably causing a *de novo* effect on a single gene. I think the fact that we are seeing it more in females than males will make sense in a minute.

If you were to take away anything from my presentation, I would say this is the message. I would say there is a group of mutations that are identified, chromosomal, single gene, in autistic patients. We do see what their primary defect is. It is a strong genetic effect, it is a very highly penetrant effect. This group tends to be dysmorphic and they tend to have cognitive impairment.

From what we know about how we found these, we know that our ascertainment method is pretty terrible, so I am speculating that there has got to be more of these which are likely, more of the same mutations. That would give you then a residual, very small group of females who are the pink section up here, and this huge group of males who are less impaired, less dysmorphic, and more puzzling as to their etiology.

The reason the paternal age effect makes sense is that we know that paternal age effects will be relevant to this group of mutations up here. For almost all of these we have equal male–female distribution. So I think there is a big chunk of autism which we maybe would have said was 10 percent or 5 percent 5 years ago, that I think is going to be closer to 40 or 50 percent of the total. That leads to the second phenomenon down here, which seems very different.

Geneticists think about things being heritable. You have heard comments about monozygous twins. I just want to make the point that *de novo* genetic events are highly heritable in genetic terminology. That is, if you take Down syndrome and you have identical twins, they will both have Down syndrome 100 percent of the time. So their phenotype of Down syndrome is determined by their phenotype, and we say the heritability is 100 percent. But their parents don't have Down syndrome. The abnormality is not inherited.

I think this is the case for all of the autism genetic defects that we know about at present. They are by and large *de novo* genetic events. We would expect them to be highly concordant in monozygous twins and much less frequently concordant in dizygous twins, which is what the bulk of the twin data says about autism.

The rest of it, I have to say, turns more to this leftover group that we understand less. I have worked with a couple of disorders, Prader-Willi syndrome and Angelman syndrome, that involve the phenomenon of genomic imprinting that I don't have time to go into here.

On chromosome 15, if you have a deletion of a particular region, you have Prader-Willi syndrome, and if you have a deletion in the same region on the maternal chromosome you have Angelman syndrome. If you inherit two copies of chromosome 15 from your mother you have Prader-Willi syndrome, and two copies of chromosome 15 from your father you have Angelman syndrome.

The deletions are genetic. If you sequence the genome, you will find 5 or 6 million base pairs of DNA have been lost. These are epigenetic. If you sequence the gene in the epigenetic cases, the sequence is perfectly normal, but the fact that these genes behave differently whether they are of maternal origin or paternal origin explains the problem.

Both of these events, which are the bulk of events that cause this kind of abnormality, are *de novo*. That is, the parents don't have the deletions and they don't have the two copies of a chromosome from a single parent.

This is emphasized for us that a diagnosis could be quite hard to figure out. If you have some cases in the mix being epigenetic, some being genetic, the epigenetic or genetic events could be *de novo* or inherited. It creates quite a complicated model, and so we have tried to explore this mostly as it relates to the patients who we understand less.

This *de novo* component fits very well with what I have been talking about, the genetic group. We don't really have definitive evidence as to whether the epigenetic component is going to be an important one or not.

In this, I am very interested, particularly from the epigenetic status, about the environmental interaction, particularly folic acid. The genotype has to have a certain epigenetic state in order to give rise to the phenotype, so I am very supportive of the idea that there will be environmental genetic interactions going on.

This graph is widely talked about and looked at. I just want to make the point that some people think there is a substantially increased incidence of autism. I think it is clear this is partly artifactual by how children are diagnosed and ascertainment and so on. But if there is any component of this that is real, it is very, very important to detect for the reasons that have already been stated this morning in terms of understanding the causation and trying to develop treatment.

If there is something going on, what could be going on? I just want to mention two issues. Prenatal ultrasound, in the event that it might not get mentioned otherwise. Paternal age we have already talked about. I want to talk about folic acid a bit more.

This is a paper from last year in *PNAS* looking at ultrasound exposure of mice and the effect on the neuronal migration in these developing mice. I think this is a good example of the kind of area that we need to be thinking about as far as any kind of environmental factor. These people made some recommendations that we shouldn't be doing prenatal ultrasound as a recreational activity, and that we need more research in this area.

This is how prenatal ultrasound has increased over the right year interval. You have seen this figure before about folic acid, and I will use it now to transition to folic acid, just to make a few points.

When I have tried to express some concern that folic acid could be a problem that could be increasing the incidence of autism, people have said it was that the fortification came too late. But I think if you look at the data, that is not correct. In the NHANES studies in the 1970s, we had 23 percent of people reporting they took a daily vitamin. In the later 1970s and 1980s, 35 percent, and this went up with time. The FDA (Food and Drug Administration) prohibited putting much folic acid in vitamins until the mid-70s, and most vitamins had none, but a few had a tiny amount. But in 1973 they raised the limit to 0.4 milligrams of folic acid. So one-a-day vitamins went from none to 0.4 milligrams in 1976 and Vidaylin went from none to 0.4 milligrams in 1977. This is before the neural tube defect perspective.

There are data from the Framingham study that people who reported that they took a daily vitamin or ate ready-to-eat cereal had a folic acid level roughly two to three times higher than people who reported they did not. This was in the 1990s before fortification. We had two groups in the population, those who were taking a folate supplement and those who were not.

So I think these changes are reasonably compatible with the possibility that timewise, folic acid is a potential factor.

Why have we been very interested in it? My laboratory has been interested in epigenetics, and it is known that using folic acid intake in mice and in humans, you can alter gene expression because of the way it contributes to DNA methylation and histone methylation.

This is a publication from some time back, where coat color in these mice is under a particular genetic element which is responsive to DNA methylation. You can change the coat color of the mice by feeding the mother differing amounts of folic acid and other methylation-related compounds during the pregnancy.

This is a study from humans. I won't try to take you through the technology, but just to say it demonstrates that folic acid can change gene expression in humans as well. This is a gene which should have only one of the two bands here present in a normal situation. These are patients with renal failure in high homocysteines, and those with the highest homocysteines are expressing both the maternal and paternal copy of the gene, so they have two bands; that is abnormal. But when you put them on folate supplementation they go back to expressing just one band, which is the normal state.

So again, folic acid can influence gene expression in mice and in humans in certain situations, and often this involves this phenomenon of genomic imprinting, where the maternal copy of a gene and the paternal copy can differ.

So folic acid definitely changes the action of some genes, probably especially imprinted genes. The laboratory acid intake of the population at large, and particularly reproductive-age women, has dramatically increased over the last three decades. Your folate level and maybe imprinted gene expression are different today than they were 15 years ago, and we need to know more about whether folic acid intake is increasing or decreasing the intake of any diseases.

The following are suggestions for potential research areas. I think genomewide studies at the exxon level and single-gene level and single-nucleotide level will expand this group, which I propose will turn out to be genetic, but we don't have very good ability to detect them right now. This will separate out this strong mutation group from the other puzzling group that is left. I think that epigenetic approaches are very worthwhile for the idiopathic portion that has not got specific genetic lesions.

Dr. Levitt: We have time for one or two clarifying questions.

Dr. Pessah: When I started out in looking at autism many years ago, only about 10 years ago actually, the emphatic view was 90 to 95 percent heritable genetics. What has changed over the last 10 years to make it 50-50?

Dr. Beaudet: Well, I don't know. It is different opinions about what the heritability is. I think one question had been, why is it so concordant in monozygote twins and nonconcordant in dizygote twins? That is totally explained by *de novo* events. Whether it is advanced paternal age causing a point mutation or whether it is trisomy 21, these *de novo* copy number variants, they all will give you 100 percent concordance in monozygote twins and a much lower concordance in dizygote twins.

I think also, this whole issue that these people have genetic conditions, their genotype determines their phenotype, but it is not inherited. So if you try to compare their genotype to their parents' genotype, you don't find the expected implications.

Dr. Herbert: You said that the known genetic mutations had a 100 percent concordance monozygotic and 5 percent dizygotic. Where are those data from?

Dr. Beaudet: I would say on general principles, if you take any new mutation event, whether it is trisomy 21 or whether it is achondroplasia, Rett syndrome, any new mutation event happens prior to fertilization or prior to twinning. The monozygote twinning takes place later on, and the twins have the identical genotype, including the genotypic error that they have.

Dr. Herbert: You are saying they are concordant for autism. We don't know whether the gene causes it or it is just a risk factor. If it is just a risk factor, then you can't assume that it is going to be 100 percent concordance.

So if it is 100 percent concordance in these genetic errors, are there data that support what you said?

Dr. Beaudet: The question is, when you find these kinds of errors, how convinced are you that they are the cause of that child's abnormality?

Dr. Herbert: Close to a risk factor, a high risk factor.

Dr. Beaudet: Right, or totally irrelevant. I think that there is some of that. If you look for these *de novo* events in the control population, you do see some *de novo* events in the normal population. But statistically, most of these new events are almost certainly the cause of the child's disability.

They all have major effects, for the most part. There may be weaker effects that we haven't discovered yet, but the ones that we are looking at here, they mostly have physical abnormalities associated with them in terms of dysmorphic features and birth defects, they are mostly mentally retarded, and they meet the criteria for autism.

Dr. Levitt: That's it. We have a lot of time for discussion this afternoon, and the godfather is looking at me. Thanks very much, Art.

Mark Nobel is our next speaker. He is going to talk about environmental factors impacting cell function. Mark is a professor of genetics at the University of Rochester Medical Center.

HOW MAY ENVIRONMENTAL FACTORS IMPACT POTENTIAL CELL-BASED MECHANISMS?[11]

Dr. Mark Noble

Dr. Noble: Thank you so much for this opportunity to come and learn from you all. It is very exciting to me to have the opportunity to take part in this discussion.

I am going to approach this talk from the perspective of our efforts to develop a comprehensive approach to the field of stem cell medicine. This work began with our initial isolation of CNS progenitor cells almost 25 years ago, and now extends to cover many components of stem cell medicine that are separate from the use of cell transplantation to repair damaged tissue.

In order to discuss our work, I have to introduce you to some of the cellular players in the CNS. The only point that I want to make with this slide is that when people talk about development, they mostly talk about stem cells and they talk about differentiated cell types, neurons and myelin-forming oligodendrocytes and astrocytes. From our attempts to understand the cellular basis of developmental maladies, however, it seems the most interesting cells are the progenitor cells that lie in the middle. These lineage-restricted progenitor cells are the workhorses of building tissues. They are the ones that are responding to environmental signals. They are the ones that are building your whole nervous system during development, and these are the ones where we focus our attention. There are several such progenitors that we study.

Our greatest interest, however, has been studying myelination because of damage to myelin being the largest category of neurological disorder, showing up in all traumatic injuries, most chronic degenerative conditions, and in respect to this meeting, with some very interesting findings in respect to autism.

Through our studies of all these progenitors and what happens to them during development, we have come to realize that many developmental maladies are diseases of precursor cells. You have abnormalities in the generation of specific cells, with specific cell types being generated too early in some conditions, and not at all in others. Or you don't make enough of certain cell types. We have been trying to

[11]Throughout Dr. Noble's presentation, he may refer to slides that can be found online at http://www.iom.edu/?id=42465.

understand how these abnormalities occur to identify the underlying principles at work when bad things happen to good cells.

There are several insights from our work that I would like to share with you. Development is a progression, in which different cells appear at different time points. In some tissues, in some lineages, they go into a single end-stage cell, in some you are generating different cell types, such as different neurons, at different times. In the specific context of myelination you have a sequence of cells, from a stem cell to the first level of restricted progenitor, to the second level of restricted progenitor, and finally on to an oligodendrocyte. There are multiple insults that we have discovered, such as thyroid hormone deficiency and iron deficiency, that can have the same outcome of not generating enough myelin.

We've been investigating the effects of different insults on aberrant myelination in a variety of ways, but the one that currently looks most interesting emerges from studies on the very ancient evolutionary problem of controlling the balance between self-renewal and differentiation. This is a central problem in understanding normal and abnormal development, and it is also an old problem, dating back to the first organisms that had multiple cell types. Our hope was that if we could solve this problem once, we might have solutions that apply to many different cell types.

We started this work as many others have done, which is to discover molecules that impact on the balance between division and differentiation. In the years of this work, we and others discovered that thyroid hormone promotes differentiation. We discovered that platelet-derived growth factor is a basal mitogen that is needed and is sufficient to promote division of oligodendrocyte progenitor cells and allows oligodendrocyte generation. We discovered molecules that suppress differentiation, such as neurotropin-3 and FGF, which enhance self-renewal. We and others also identified other factors, such as bone morphogenetic protein, that enhance astrocyte generation.

What we then wanted to know is how all this different information received by a cell becomes integrated. Lots of people study this integration question, and we see this diagram of intersecting signaling pathways in many meetings. But we wanted to ask a different question, which is, how does this information become integrated in the context of physiology?

The physiology that we have been most interested in is redox physiology. I want to explicitly say that I am not that interested in oxidative stress. What I am interested in is the normal use of redox balance in controlling development and cell function. This is a different question from that asking about stresses that kill cells. We study oxidative stress a bit, but what has turned out to be much more interesting is this area of normal physiological control, which is simply the balance between reducing and oxidizing equivalents in a cell.

The first discovery we made in this area came from studies in cell death. What we found is that modifying the redox state by tiny amounts, by 15 percent, has tremendous effects on biological outcomes. For example, here we are looking at survival of spinal ganglion neurons, given suboptimal amounts of nerve growth factor in the purple bars here. If we add *N*-acetylcysteine, a cysteine pro drug that is taken up by cells and can be used to increase glutathione levels, we can give enough *N*-acetylcysteine to cause a 15 percent change in glutathione content, and that is sufficient to obtain a 1,300 percent increase in the number of surviving neurons.

We wanted to understand how it was possible for small changes in redox state to have such large effects on cellular function, and we started taking this apart through our work on progenitor cells. Through our work on progenitor cells, we discovered a number of general principles.

It now is clear that the redox state is a central regulator of precursor cell function, controlling whether a cell divides or differentiates as well as whether cells survive. For example, we found that all the classic signaling molecules that we study converge on regulation of redox state. Neurotropin-3, FGF, thyroid hormone, BMP, every signaling molecule we have examined converges on redox state, and the redox changes that they induce are necessary for them to exert their functions. We have also discovered that the organism uses developmental genetic or redox state to control precursor cell function.

There are some general principles relevant to our data that I want to mention. If an oligodendrocyte progenitor cell is a little more reduced, it is more responsive to mitogens and survival factors. If it is a little more oxidized, it is more responsive to differentiation and death factors.

Now let me tie it into some of the discussions that have occurred this morning. We decided to start working in the field of toxicology because we wanted to study redox perturbations that have real-world significance. In the laboratory we carry out genetic manipulations and pharmacological manipulations, but it is toxicology that has real significance for all of us in the world.

Many of these toxicants are pro-oxidants. So we started working on them. We started with methylmercury, for which there is an extensive literature. This literature says that astrocytes and neurons have an LD-50 for methylmercury of about 500 nanomolar, equivalent to 100 parts per billion. This is a level of methylmercury exposure that would occur only rarely.

When we went to study progenitor cells, we found that the ideas of vulnerability of cells of the CNS were entirely off base, and that progenitor cells are vulnerable to methylmercury at concentrations of 20–30 nanomolar, that is, from 4 to 6 parts per billion. It currently appears that all the progenitor cells that we look at are vulnerable to things like methylmercury and thimerosal and other things at these exposure levels.

What we found, which was recently published in *PLoS Biology*, is that environmentally relevant levels of toxicants make cells more oxidized precisely in the range of relevance to our work on development and on cell-extrinsic signaling molecules. And, just as we would predict, progenitor cells exposed to such pro-oxidants drop out of division and become more vulnerable to other physiological stressors.

What we did next was to take apart signaling in the cell from the nucleus back to the receptor to understand the mechanistic basis by which toxicants disrupt normal progenitor cell function. What we discovered when we looked first at the PDGF pathway was that the effects were absolutely confusing, as our signaling colleagues told us they would be and the literature told us they would be, which was that everything was suppressed—NF kappa B-mediated transcription, serum response element-mediated transcription, Erk phosphorylation, AKT phosphorylation, and phosphorylation of PDGF receptor.

We next were fortunate to choose the right control experiments to carry out, which was to look at the effects of methylmercury exposure on signaling pathway activation by other ligands. What we next asked was what happens to neurotropin-3 (NT-3) signaling in these cells. The answer was that environmental toxicants had no effect whatsoever. So that suggested that there is specificity in these changes and that specificity resides at the receptor level.

By looking at the PDGF signaling pathway and the NT-3 signaling pathway, we found a novel regulatory pathway that appears to be a convergent point for multiple chemically diverse toxicants. What happens when a cell becomes oxidized, apparently regardless of the oxidizing agent that is chosen, is that the cell activates Fyn kinase, an enzyme that is a member of the src family of kinases. Activated Fyn then activates a ubiquitin ligase called c-Cbl. c-Cbl is a negative regulator of receptor signaling for some receptor tyrosine kinases. What happens when you activate this pathway is that c-Cbl attaches ubiquitin to the receptor for platelet-derived growth factor, the EGF receptor, C-Met, and some other receptors of interest, and they become degraded more rapidly. As a consequence, you suppress all the downstream signaling because you don't have as many of the receptors anymore.

Methylmercury has no effect on trkC (the receptor for NT-3) because this receptor is not a c-Cbl target. In fact, receptors that are not c-Cbl targets completely appear to be unaffected by activation of the redox/Fyn/c-Cbl pathway.

Our current studies demonstrate that multiple substances with pro-oxidant activity converge on Fyn activation, including multiple toxicants, multiple chemotherapeutic agents, ethanol, thyroid hormone, and other agents. This convergence may be due to activation of Fyn kinase by oxidized glutathione. So if you increase oxidized glutathione, you activate this pathway.

We invested many years in developing in vitro systems that mimic what happens in the animal, so that we could use in vitro studies to accurately predict in vivo outcomes. Indeed, that is the case for our studies on toxicology.

If we expose developing rats to levels of methylmercury in ranges as low as 100 parts per billion in the maternal drinking water, levels that are 10 percent or less of what other people mostly study, what we see in vivo is that cerebellum levels of PDGF receptor are decreased, levels of epidermal growth factor receptor are decreased, and levels of neurotropin-3 receptor are not affected. The same is true in the hippocampus. Consistent with a lot of our other work and redox regulation in respect to development, the cortex is unaffected, just as we would predict. It is very interesting to us that the cerebellum is a target in so many toxicant exposure paradigms, and is of interest of course in autism.

We next want to understand why do some people have an outcome that is bad and other people have no outcome at all that you can see when they are exposed to similar levels of environmental toxicants. So we got very interested in strain differences. The literature taught us that SJL animals are responsive to organic mercurials and CBA animals are not. The literature also says that this is because of immune system problems. We were hoping that maybe that wasn't the case, so we purified progenitor cells and asked at the single cell level what happens.

It turns out that progenitor cells from the corpus callosum of SJL animals are much more vulnerable to anything that we throw at them. For example, here we are looking at cell division in the corpus callosum following a clinical exposure of thimerosal. You have reduced division in the corpus callosum and subventricular zones of SJL animals, but not in CBA animals. We have similar effects in tissue cultures of progenitors from these animals, where cells are more vulnerable in a strain-dependent manner.

Our current hypothesis is that SJL cells are more vulnerable because they are more oxidized, and indeed they are more oxidized. As we look at other strains of mice, what we are finding is that if the cells are derived from the animals that are more oxidized, they are more vulnerable, and if they are derived from animals that are more reduced, they are more resistant.

We next asked whether, if you have a genetically more oxidized animal, whether this Fyn/Cbl pathway itself is more activated just because you are genetically more oxidized. That would be the outcome of our predictions, and indeed, it is what happens. If we look in the central nervous system of SJL mice, they have lower levels of PDGF receptor than we find in CBAs. And they have lower levels of the C-Met receptor. So it looks like being genetically more oxidized also activates this pathway. Thus, we are currently thinking that being more oxidized, which occurs frequently in children with autism spectrum disorders, is a marker (and potentially a mechanism) of vulnerability that needs to be more closely studied.

Finally, because I mentioned thimerosal, I want to be explicit about what I think our thimerosal results say at this point in time. It is actually quite remarkable to find how little data exists on thimerosal toxicity. In our studies, we do not see any difference between thimerosal and methylmercury. Thimerosal is as toxic as methylmercury, but it also is not more toxic than methylmercury. Based upon these biological discoveries, it is appears that it is not wrong to be concerned about the possible contributions of thimerosal to neurological syndromes. But I would say that such a concern needs to be studied in the context of the idea that there is a subpopulation of children more vulnerable to the effects of exposure to potentially toxic substances, thus, if there is something particularly interesting about thimerosal. I think what it may be is that you have susceptible individuals, and the probability of a susceptible individual being exposed to a high level of thimerosal would have been higher than if he or she might be exposed to methylmercury or lead simply because it was used in so many vaccines in the 1990s. That is what would make thimerosal unique, rather than its chemistry. At a cellular level, it is important to stress that all the environmental toxicants that make cells more oxidized are something to be concerned about as potential disruptors of normal nervous system development.

Dr. Levitt: We have time for a few clarifying questions if anyone has any.

Dr. Hertz-Picciotto: I am just wondering if you might comment on how these results might apply to the very consistent finding of Purkinje cell loss that has been seen in so many studies of autism.

Dr. Noble: We have worked out the effects of early mercury exposure in SJL versus CBA animals. In CBAs we see no effect on Purkinje cells; in SJLs it is a disaster. Cell membrane is reduced, they are out of position, the carburization is compromised. So yes, we see a very strong correlation.

Dr. Levitt: Any other clarifying questions? We have a lot of general discussion time.

Dr. Pessah: I just have a question about the last 30 years of neuroscience and the dish. I'm not sure that even 10 percent of those that study cells in the dish, including progenitor cell progression, are not using 20 percent oxygen in their incubator. So essentially everything is underoxidized when you make the measurement.

Dr. Noble: Not everyone, and not us. The question, so that everybody knows what we are talking about, when most people do tissue culture experiments they are growing cells in atmospheric oxygen. The baby is exposed to an oxygen concentration, depending upon which papers you believe, of somewhere between 3 percent and 5 percent. Atmospheric oxygen is 21 percent. That would oxidize cells and make them do strange things.

We have looked at this extensively. What we find is that for most parameters that we study, the cells don't actually change their behavior between 5 percent and 21 percent. We continue to run them at 5 percent because we believe they should change their behavior, but it looks as though—if one wants to talk about this on an evolutionary basis, the idea would be that there is no selective pressure in the brain to respond to 21 percent oxygen. It never sees it. It is not a relevant situation.

In contrast, the difference between exposing cells to 5 percent and 1 percent, a hypoxic insult, is enormous. So we know we can change cell function by changing oxygen tension, but remarkably, the cells have been very, very resistant to 21 percent, even though we continue to run everything at 5 percent.

Dr. Levitt: Thanks very much, Mark. We have one more presentation before the break. Theodore Slotkin is going to talk about animal models, to take us from cells to animals. Dr. Slotkin is professor of pharmacology and cancer biology at Duke.

HOW MAY ANIMAL MODELS BE USED TO EXAMINE POTENTIAL ENVIRONMENTAL-BASED MECHANISMS?[12]

Dr. Theodore Slotkin

Dr. Slotkin: When we are dealing with autism, humans are the animal models of choice. If we could draw a connection between specific developmental neurotoxicants and ASD or autism per se, what we would like to be able to do is to use animals to define prototypes, that is, types of compounds that define entire families that attack the developing brain in the appropriate manner, that would then enable us to guide human investigations as to what we might look for in human populations with ASD as potential causative links. Then in the reverse direction, we could take things that people had noticed from clinical or epidemiological studies and then do the animal studies that could potentially prove cause and effect. The thimerosal story would be traditionally one of those where you go from suspected human exposures back into the animal.

But the real impediment is that there actually is no animal model that gives us a complete picture of ASD. Therefore, what we have to concentrate on at this stage is to identify the mechanisms that contribute to adverse neurodevelopmental outcomes, which may include ASD, but are not restricted just to ASD, to point out the types of chemicals we might want to be concerned about and thereby guide future clinical

[12]Throughout Dr. Slotkin's presentation, he may refer to slides that can be found online at http://www.iom.edu/?id=42466.

investigations.

We are all awash in a sea of tens of thousands of neuroactive chemicals. That raises some very serious questions that I think animal models help us address. That is, first of all, why is it that we adults are awash in them and our brains aren't permanently affected, whereas the developing organism is? What is it about development that renders neuroactive agents capable of producing permanent alterations? Why is there a critical period for it? Why is it that exposure before or after a certain stage doesn't do what exposure during a critical period does?

A key question was just addressed in the previous talk: Why do apparently unrelated agents produce similar outcomes? I am going to show you that that is not just restricted to the issue of oxidative status inside the cells, but there are very cogent reasons for a critical exposure period, potentially even more important than what that compound is.

Then finally, I am going to show you an example of this trade-off of how animal studies can help guide human studies for studies of autism using an example of work we were doing that we originally thought was totally on a different topic, using the drug terbutaline and its action on beta-adrenergic receptors.

We will take those in sequence. Why is it that development is special and the developing brain especially vulnerable? There is a good reason for that. One of the purposes of the developing brain is to assemble itself and learn. For example, when you are young you can learn multiple languages very easily, and after that it becomes much more difficult. Cells are the same way. They have a specific period in which they use their inputs in order to learn. In other words, the brain has a certain degree of hard-wired development, but superimposed on that is input from the environment, which influences the subtle but important connections that define us as individuals and that ultimately impact on all our important behaviors.

One of the ways in which environmental input is transduced into a change in the development of neural cells is the release and actions of the same small molecules that communicate ordinarily across the synapse, that is, neurotransmitters. But during development, neurotransmitters aren't just neurotransmitters. They are trophins that influence the fate of their target cells. They do so through the very same receptors that our adult brains use for ordinary synaptic communication, and the same signaling cascades. The difference is that during development, the actions of these molecules change the repertoire of genes that are read out of the nucleus in the target cell, and depending on the stage of development of the cell, those genes might be involved in the control of cell division or differentiation or growth or apoptosis, since programmed cell death is a part of the modeling of the brain, or what we will be concerned about here—learning.

There is a specific period in which input to a developing neuron teaches the neuron how to respond to specific inputs, both during development and for the future, life-long function of that particular circuit. The critical thing about this is that the same neurotransmitter and the same receptors and the same signaling cascade can do all of these things in a given cell, depending on the stage of development of the cell. And each cell doesn't just have one neurotransmitter, it has multiple ones.

So what that says is that if cells learn during a critical period, learn in the same way we learn language but at the cellular level, then when you provide an input during that period, you change the fate of the cell, and that changes what it is going to do permanently. Whereas, for our mature brains, input after the critical period just produces short-term responses, and if we continue to try to elicit the response, we become desensitized to that particular input—it gets ignored in the same way that a continual sound eventually disappears from active perception. It is subject to short-term, reversible compensatory adjustments.

Let me show you that as an illustration of a presynaptic nerve terminal and its postsynaptic target cell in the mature brain. If you reduce the amount of neurotransmitter in the synaptic cleft, this cell will respond by increasing the number of receptors and augmenting the ability of those receptors to produce a response. So you have small input, the cell is sensitized so as to maintain the response; the reduced input is offset by enhanced responsiveness.

If you overstimulate this cell and make a lot of neurotransmitter appear in the synapse, then the cell will desensitize. It will lose its receptors from the surface and it will uncouple them from the ability to elicit a response inside the cell. So this is a negative feedback loop. A small amount of input results in a boost in the ability to elicit a response so that you can make that synapse work. Stimulate the synapse too much, and the postsynaptic site shuts off and terminates the signal.

In contrast, during development, because these synapses are learning, that relationship is reversed. A lot of stimulation during development promotes the development of the appropriate receptor for that neurotransmitter, and therefore augments the response. This is a positive feedback, so that is a good thing for learning. A little bit of stimulation teaches. The good side is that this is what enables you to learn, or as I will now illustrate, even survive getting born. For this illustration, I am going to concentrate on peripheral responses, because I am going to come back to this at the end for messages for the future.

Stress hormones, especially adrenaline, rise precipitously at birth, and if this were the adult you would expect to see desensitization of the responses for that transmitter. Instead, in the developing organism, that response increases during the period in which the stimulus is rising. Now, if I use terbutaline, a stimulatory drug that acts like adrenaline on its target receptors—the beta-adrenergic receptors—in the adult that would desensitize the response, but in the newborn it instead sensitizes the response. It teaches the cell: Beta-receptor stimulation in the immature organism leads to increased expression of the receptors and increased coupling of receptors to responses.

Now, what if we block the stimulus? If, for example, we destroy the nerves that are supplying that response, then the cells will never learn how to respond to this particular input. In this illustration, you can see the normal response in rats at ages ranging from adolescence to adulthood: the early denial of input results in permanent loss of responsiveness. In other words, there is a critical period in which you must teach those cells what their fate is going to be—this is learning at the cellular level.

The same principles operate in the developing brain. They are universal for all neural circuits, not just peripheral systems. This input–output relationship originated very early in evolution and in fact, they can be seen in lower organisms such as sea urchins and bacteria, where they use the same molecules and receptors as a way of controlling their own growth and development.

Thus, if neural input programs response development, that means there is always a critical period in which any disruption of input, positive or negative, will permanently alter function. If you send the wrong signal, it is going to produce the wrong outcome. It is going to teach the cell to respond incorrectly to the wrong kind of input. It will make that change permanently.

What kinds of things will do that? Any drug or chemical that is neuroactive works ultimately by reinforcing or blocking the actions of a neurotransmitter. This will mean that exposure to drugs of abuse or therapeutic agents that are neuroactive, environmental contaminants, whether they are organometals, which influence things like signaling and oxidative stress, insecticides, all of these can lead to misprogramming of responsiveness.

Let me give you an example of this with terbutaline and its potential connection to ASD. Terbutaline is a drug that is commonly misused in preterm labor. It is a beta-adrenergic stimulant, so it works the way adrenaline does. It inhibits uterine contractions, but it is effective only for about 48 to 72 hours in doing that. Then as you would expect, in the adult it desensitizes the response so that preterm delivery proceeds anyway. Accordingly, it is not supposed to be used for maintenance tocolysis, but nevertheless it still is frequently used for that purpose.

We were doing studies on this agent in the 1980s and 1990s, primarily to study receptor regulation. As predicted, terbutaline produces permanent changes in responsiveness because it acts just like overstimulating the nerves supplying those receptors.

Coincidentally with that, there were some clinical and epidemiological findings indicating that maintenance tocolysis with terbutaline was resulting in adverse neuropsychiatric outcomes.

In the last decade, we performed definitive studies to show that terbutaline is indeed a developmental neurotoxicant that affects specific targets in the brain, notably cerebellar Purkinje cells, areas of the hippocampus and cerebral cortex. For our purposes, though, the important finding is that terbutaline treatment in rodents shares morphological and behavioral characteristics that are found in autism. It does involve things like oxidative stress and neuroinflammation. There is a critical period corresponding to the second trimester of brain development as it would be in the human fetus. There are also peripheral changes in cardiovascular reactivity and other peripheral functions that are similar to those that are reported to act in autism.

Consequent to our studies in animals, which were not originally conceived with a relationship to autism, a study done by Dr. Andrew Zimmerman's group and others pursued the connection between the use of terbutaline in preterm labor and the incidence of autism. They found that if you gave terbutaline for 2 weeks or longer, there was an increased risk of autism spectrum disorders. If you superimposed that on a receptor polymorphism for the beta-2-adrenergic receptor that prevents the receptor from desensitizing, the risk becomes much higher.

They concluded that prenatal overstimulation of beta-adrenergic receptors by terbutaline by itself or in combination with these genetic polymorphisms were responsible for autism in this particular cohort. What I want to point out is that that is an example of how animal models can lead to unsuspected mechanisms that can be responsible for neurodevelopmental disorders, an approach that is not strictly limited to autism.

How does this point out a path for future research efforts? First, rather than relying on accidental connections, the way our two laboratories did for the case of terbutaline, there is a lot of information available on developmental neurotoxicants from the animal literature and databases that could point to retrospective examinations relating these exposures to ASD.

On a population level, even GIS information might prove useful for examining how having been brought up in an area where, for example, organophosphate pesticides were used heavily, given that these agents are suspected as contributors to the increased incidence of ASD. Terbutaline and organophosphates are potential examples of a rational approach to using existing findings to search out new connections between exposure to chemicals that may be contributing to the rise in ASD. As we just heard, agents that produce oxidative stress are also good candidates for this database or literature approach to trying to draw mechanistic connections from existing information.

Our study showed that the results from animals can be used to trigger the comparable studies of human populations for exposures and outcomes. I think there is an added value from examining an outcome where you can readily document the exposure. Terbutaline is different from most environmental exposures in that regard because the exposure is documented in an individual's medical records. But there are certainly many other compounds or classes of compounds that can provide similar types of leads for future investigation.

The basic problem is that we have potentially as many as 50,000 chemicals we have to screen for developmental neurotoxicity, so we are not going to be able to proceed one at a time, or at least not in animal models. We are going to need high-throughput models. As just one approach, cell culture models for neurodevelopment are plentiful and easy to use. We can then use all of the armamentarium of siRNA or humanized cells to explore the role of specific genes or genetic polymorphisms that are found in autistic human populations, and insert them into cells and see what they do to vulnerability to developmental neurotoxicants.

There are efforts going on at NIEHS as well as in our laboratories and elsewhere to use lower organisms as ways of screening for these. As I pointed out, the peripheral surrogates that are predicted by common outcomes for autonomic input from the central nervous system to the periphery might then be an additional role for future use of studies.

To sum up, the big impediment here is that there are too many candidate molecules to study, too little time, and too little money, and consequently designing high-throughput screens for developmental neurotoxicity is a mandatory first step to drawing connections between environmental chemical exposures and ASD as well as other neurodevelopmental disorders.

Dr. Noble: I would just like to reinforce that. Our work shows that it is trivial to screen chemicals. With these high-sensitivity systems that we have, we have multiple parameters now that let us analyze things as fast as we can grow dishes of cells, frankly.

It is very, very simple, so long as you use the right cells. It is very critical. If you use established cell lines it is pointless because they have mechanisms that protect them against these kinds of insults. You have to use early progenitor cells to get high sensitivity to outcomes.

A lot of the biomarkers that are coming out from this look pretty intriguing, particularly when I look at the studies from Dr. James. The changes in metabolic components of oxidative state that are being reported in autism are very much like things that we are seeing. We also know how these proteomic markers that we are finding very reliably, some of which are completely independent of transcriptional changes. So when we see changes in receptor levels, there are no transcriptional changes at all.

So I think we are building up between the transcriptional work, the protein work, the metabolic work, we are building up the tool kit. So we need to do this.

Dr. Schwartz: In the human study that you showed us, you showed us a relative risk of 2.4 and 4.4, but what percentage of the cases was it associated with?

Dr. Slotkin: That is not my study. I'm just citing the literature. I'm not familiar with the details of it.

Dr. Schwartz: And in your own studies, did you see any strain differences in responsiveness?

Dr. Slotkin: We were using outbred rats, so strain differences is a moot point.

Participant: It might be very naive, but are there any natural animal models of autism? And if not, why not? And if there are, wouldn't you expect epigenetic environmental factors to impact on that?

Dr. Levitt: Why don't we save that for the general discussion? Our next speaker is Ira Lipkin. He is the director of the Greene Infectious Disease Laboratory and other major titles at Columbia University. He is going to be talking to us about infection and immunity in autism.

AUTISM, INFECTION, AND IMMUNITY: WHAT ARE THE POTENTIAL CAUSATIVE ENVIRONMENTAL FACTORS AND HOW CAN THEY BE IDENTIFIED AND PRIORITIZED?[13]

Dr. W. Ian Lipkin

Dr. Lipkin: My task is to describe technologies that might be applied to answer questions relating to infection and immunity in neurodevelopmental disorders. We like to think of these technologies as "peace dividends" because they were developed with support from biodefense funding that came online after 9/11.

To set the stage for consideration of how diagnostic and surveillance technologies can be implemented, I must first discuss the mechanisms by which infectious agents can cause disease. This introduction is critical because if infectious agents play a role in pathogenesis of autism spectrum disorders, common conceptions of mechanism may not apply.

We typically think of infections as associated with acute illness at the sites where the agents replicate. Poliovirus, for example, is an enterovirus that causes gastrointestinal dysfunction in many people, meningitis in a smaller subset, and paralysis due to death of motor neurons in brain and spinal cord in a still smaller subset. Vibrio cholera infects the intestine to cause diarrhea known as "cholera." However, less obvious relationships may also be important. Clostridium botulinum grows in the skin or the gastrointestinal tract, and elaborates a toxin that travels systemically to cause paralysis through a remote effect at the neuromuscular junction. There are also instances, such as infection with hepatitis B or hepatitis C,

[13]Throughout Dr. Lipkin's presentation, he may refer to slides that can be found online at http://www.iom.edu/?id=42467.

where the agent itself has only a modest direct effect; however, immune responses to infection and the associated inflammation results either in death of cells and organ failure, or neoplasia. Some microbes can cause immunosuppression, resulting in disease due to infection with other, frequently opportunistic organisms. In our era we think of HIV as the prototype for this scenario; however, virus-induced immunosuppression was first described by von Pirquet in the 1800s in the context of measles. Persistent viral infections can have subtle effects on differentiated cell functions such as the capacity to make an enzyme or a neurotransmitter. Although this mechanism has not yet been shown in humans, its plausibility has been demonstrated in animal models of persistent viral infection resulting in dementia, type 1 diabetes mellitus, and hypothyroidism. Molecular mimicry, a mechanism by which an antibody or T-cell mediated immune response against a microbe results in damage due to cross reactivity to a normal host component, was described by Sue Swedo this morning. We must also consider effects of infection that are locked into specific windows of vulnerability. This is particularly pertinent in autism where such windows have been defined vis-à-vis exposure to thalidomide. Ezra Susser highlighted related examples wherein risk of schizophrenia was increased by prenatal stressors, including infection. The hope, of course, is that discovery of agents that cause disease by any of these mechanisms will facilitate the development of preventive and/or therapeutic strategies that promote public health. A major recent advance along these lines was the development and approval of a vaccine for papillomaviruses that is anticipated to have a profound effect on the incidence of cervical cancer. Whether we will be able to achieve similar success in autism remains to be seen; however, I think that is highly unlikely. My view is that no single agent or group of agents will be implicated. Instead, to the extent that environmental factors are important in pathogenesis, we will discover that toxins, infection, and stressors of various types, can activate similar pathways to cause similar effects.

Finally, although I have been asked to focus on the role of pathogens, I want to emphasize that our working model is one with three dimensions where genetic susceptibility, environmental triggers, and temporal context act in concert to cause disease. Thus, a comprehensive investigation must address the intersection of all three components.

This slide from the *New Yorker*, published during the West Nile virus outbreak of 1999, shows medical staff examining a pharaoh in a hospital bed in classic garb. The point is that differential diagnosis of infectious disease is rarely this straightforward. Hence, we need laboratory diagnostics.

The next several slides describe systems that together define a suite of diagnostic tools for use in differential diagnosis of infectious diseases.

The first system is MassTag PCR, a multiplex PCR platform that allows us to rapidly query 20 to 30 different infectious agents simultaneously. Throughput is rapid and assays are inexpensive. The second is the GreeneChip, an array that allows us to consider thousands of agents, but at lower throughput and higher cost. The third is metagenomic sequencing. More time-consuming and expensive yet, however, extraordinarily powerful and uniquely suited to cataloging microflora and discovering new pathogens.

MassTag PCR panels have been established for detecting the vast majority of infectious agents causing respiratory disease, hemorrhagic fever, meningitis, encephalitis, and diarrhea. In 2005, application of this method allowed us to discover a new rhinovirus.

GreeneChips are glass slides, similar to the types of slides that many of you may use for transcript profiling, that are decorated with thousands of probes. Nucleic acids in samples are amplified by PCR, and applied to slides. Binding of amplification products and of a fluorescent label allows detection of microbes. Various formats have been developed for different applications: all viruses, respiratory pathogens, all known vertebrate pathogens. Analysis has been automated such that images from slides can be submitted from remote locations and analyzed. In this instance, we identified influenza A virus in a nasopharyngeal swab; more detailed analysis allowed us to determine that it is H5N1.

Implementation of diagnostic technologies allows one to discover not only new pathogens associated with disease, but also known pathogens in asymptomatic individuals. In the context of a study of environmental triggers of type 1 diabetes mellitus in a Norwegian cohort, we recently examined stool samples from children using MassTag PCR and GreeneChips. To our surprise we found evidence of frequent infection with enterovirus subtype 71, a virus typically associated with paralytic illness in Asia. Genome sequencing and phylogenetic analysis revealed mutations in the Norwegian viruses that we predict impair replication and ability to cause disease.

Our most expensive and labor-intensive platform is metagenomic sequencing. Although applications of this platform in whole-genome sequencing are well described, its use in pathogen discovery is relatively recent. In essence, thousands of short sequences are aligned and assembled into continuous strings. Host sequences are subtracted, yielding a series of candidates that represent potential pathogens. Candidate sequences are analyzed for similarity at the nucleotide level and at the protein level for viruses, bacteria, parasites, or fungi.

Metagenomic sequencing can be employed in real time. Here is an example wherein we investigated a cluster of deaths in organ recipients linked to a single donor. No agent was identified through classical methods such as culture, serology, or PCR. Thus, tissue samples were referred to us for study. After failing with MassTag PCR and GreeneChips, we moved to metagenomic sequencing. Analysis of 140,000 sequences led to detection of a novel arenavirus. This virus was subsequently implicated through specific PCR and serological investigations.

And now—à la Monty Python—for something completely different. Several weeks ago, at another IOM conference, I met Diana Cox-Foster, an entomologist and microbiologist studying Colony Collapse Disorder (CCD). This is an extraordinary phenomenon wherein honey bees inexplicably leave their hives and don't return. Given the importance of pollination for agriculture, we began using metagenomic sequencing to examine the microflora of bees with and without CCD. This work yielded intriguing leads that may provide insights into CCD; however, the point I want to make is that metagenomic sequencing allows one to simultaneously define complex microflora in a sample: bacteria, viruses, parasites, and fungi.

We have been asked to comment on what resources might be helpful in addressing the role of the environment in autism pathogenesis. In this spirit I would like to note the Autism Birth Cohort (ABC), a prospective birth cohort based in Norway of 100,000 children and their parents, that collects biological samples and clinical data beginning at the 17th week of gestation. ABC collections, joined with the diagnostic platforms for microbiology, toxicology, and genetics, will enable new strategies for examining gene–environment–timing interactions in health and disease. Those of you interested in learning more about the ABC may wish to visit its website at www.abc.columbia.edu.

In summary, microbial pathogenesis is complex. Susceptibility is a function of genes, age, and other factors. Mechanisms can be direct or indirect. Expression of disease may be delayed. The microbiome is largely uncharted; however, new tools for microbial surveillance will change the landscape for understanding chronic as well as acute diseases.

I will close with this wonderful quote from Einstein. In the period that he was active as a professor, a student remarked, "The questions on this year's exam are the same as last year." "True," Einstein said, "but this year all the answers are different." With new models and new strategies for addressing them experimentally with clinical materials, we may shed new light on the role of environmental factors in the pathogenesis of autism and related disorders.

Dr. Levitt: We have time for some questions.

Dr. Leshner: You went through a slide very suddenly and then got off it, and I can't remember my question, but I can remember what was on the slide.

Dr. Lipkin: Neurodevelopmental disorder? Is that the slide you wanted?

Dr. Leshner: No, one more slide.

Dr. Lipkin: The next slide is this. This is the slide you want.

Dr. Leshner: This is an incredibly gigantic project which you skipped over rather rapidly.

Dr. Lipkin: That is because I was told I didn't have time.

Dr. Leshner: I know. What is this? Who is doing it? Who is paying for it? 100,000 children?

Dr. Lipkin: The National Institute of Neurological Disorders and Stroke (NINDS) is supporting it. It is called the Autism Birth Cohort. It is nested within something called the mother and child cohort. It is in Norway. There are several people who are working with this cohort at present: Ezra Susser, Deborah Hertz, Mady Hornig, Alan Wilcox, and then we have counterparts in Norway as well.

This was conceived to do the same sorts of things that all the other birth cohorts are conceived to do, but because we were later to get started, we were able to focus more on proteomics and transcriptomics and viromics and any other -omic you want to think of, so that we could try to address these kinds of questions.

Our vantage point was that it was going to be gene–environment–timing interactions. That is the principle that guides this study.

Dr. Leshner: My question is, when do you start taking samples?

Dr. Lipkin: We started taking samples already.

Dr. Leshner: No, I meant in development.

Dr. Lipkin: The first visit, which often occurs as late as 17 to 18 weeks' gestation. The mother has consented and blood is collected. Then there is blood collected from there on. So you can do serology and look for changes in titer and so forth.

Dr. Levitt: I wanted to ask one technical question. The different platforms that you used, some are qualitative and some are quantitative.

Dr. Lipkin: Those are all qualitative. They are purely surveillance tools. You have to follow them, using real time or an equival.

Dr. Levitt: So presence–absence.

Dr. Lipkin: Correct.

Participant: I had a question about the developmental disorders and children with autism. I think there is a deeper question here. Children with developmental disorders and also autism had more infections from multiple pathogens, but what about the effect of the pesticides or other neurotoxicants on responses?

Dr. Lipkin: I have no clue.

Dr. Levitt: We are going to talk about that toward the end during the discussion time. I think it is going to come up.

Dr. Schwartz: Ian, I had just one clarifying question for you. Just to be totally clear here, your microbiome is part of this 100,000 cohort study?

Dr. Lipkin: Yes.

Dr. Schwartz: And how many patients are you doing it on, what time points?

Dr. Lipkin: At present, all we have right now is the tool kit that was built for detecting West Nile virus and avian flu. But the remainder, which is the collection of these samples and so forth, that is supported, and those materials have been collected. In fact, the first samples have only recently come out of Norway to Columbia for analysis, and those are cord bloods. So we haven't done anything with those samples as yet except some of the stool samples, which I showed you. We just found a wide variety of viruses.

Dr. Levitt: Thanks very much, Ian, that's great. Our next speaker is Jill James, who is professor of pediatrics at the University of Arkansas School of Medical Sciences.

ENVIRONMENTAL FACTORS AND OXIDATIVE STRESS: HOW MAY OXIDATIVE STRESS IMPACT THE BIOLOGY OF AUTISM? WHAT FACTORS MAY BE CAUSING THIS OUTCOME?[14]

Dr. S. Jill James

Dr. James: For the next 15 minutes, I would like to explore with you pro-oxidant environmental exposures and the possible implications of redox imbalance in autism.

Let me begin by explaining that redox imbalance is actually a relative term. In fact, it is a continuum from subtle shifts in redox homeostasis up to more severe imbalance that is associated with oxidative stress and pathology.

Most reactive oxygen and nitrogen species are generated endogenously from normal oxidative metabolism. They are also generated exogenously, and most relevant to our discussion today are pro-oxidant environmental exposures. Multiple or chronic exposures to pro-oxidant environmental toxicants can sustain redox imbalance and lead to oxidative damage and promote complex disease.

Counteracting these sources of oxidative stress is a wide variety of antioxidant defense mechanisms. Chief among these is glutathione. The ratio of the reduced active form of GSH to its inactive oxidized form, GSSG, is considered to be the best indicator of intracellular redox status. A decrease in GSH and/or an increase is GSSG will negatively affect intracellular redox homeostasis.

Most research in oxidative stress is focused on the damaging effects, but I would like to make a point that is less commonly appreciated, and that is that the small subtle shifts in redox balance are in fact beneficial and represent essential signal mechanisms for normal cell function, as Dr. Nobel has elegantly demonstrated.

Redox signaling is important for cell cycle status and for the activity of a multitude of redox-sensitive enzymes. The redox status of the cysteine in the active site will activate or inactivate these enzymes. In addition, gene expression, transcription factor binding, and chromatin remodeling are affected by small changes in redox status as well as the activation of the innate immune system and the inflammatory response.

---

[14]Throughout Dr. James's presentation, she may refer to slides that can be found online at http://www.iom.edu/?id=42484.

On the other hand, chronic or severe shifts in the redox ratio can be irreversible and promote a self-perpetuating cycle of oxidative stress and damage. These include glutamate toxicity, inhibition of redox-sensitive enzymes, protein misfolding, mitochondrial dysfunction, and cell death. These more severe pathologic changes can lead to accelerated aging and contribute to the pathogenesis of complex diseases, particularly neurologic disease.

I think an important point for us to consider in the discussion today, and Dr. Nobel has referred to this as well, is that we are exposed to a wide variety of structurally different chemicals in the environment which include multiple metals, solvents, pesticides, industrial chemicals. Importantly, although they are all very different structurally and chemically, they all share a common mechanism of action, which is to induce oxidative stress and deplete glutathione. They are all pro-oxidant.

That suggests that there may be a common molecular mechanism of toxicity underlying these very different and diverse chemicals present in our environment. A common mechanism of action would certainly simplify the search for an association between autism and the environment.

Most of these chemicals, their safe levels and toxic levels, are analyzed individually, but we all know we are exposed to complex mixtures. Multiple simultaneous pro-oxidant exposures are additive and can even be synergistic in toxicity. This implies a very important point in toxicology: multiple subtoxic exposures can become toxic when they are combined.

Glutathione is not only the major intracellular antioxidant; it has important detoxification functions as well. Heavy metals have a high affinity for the sulfhydryl group on glutathione and bind spontaneously. There is a wide family of glutathione S-transferases that will enzymatically create water-soluble glutathione conjugates that are then metabolized and excreted in the bile and the urine. So glutathione is not only the body's major natural chelator, it is a major mechanism for the elimination of many of these environmental toxicants.

Resistance or vulnerability to pro-oxidant environmental exposures depends largely on intracellular glutathione levels. Depleted glutathione reserves will increase sensitivity to pro-oxidant exposures. With very robust glutathione reserves and a high GSH/GSSG ratio, toxic insults will be buffered and will never reach a toxic threshold. On the other hand, with fragile or depleted glutathione reserves, the same toxic insult can precipitate toxicity and pathology. We have recently shown that autistic children have a lower glutathione/redox ratio. We and several other investigators have started to identify metabolic biomarkers that suggest that many of these children may be under chronic oxidative stress.

This is a flow diagram of the metabolic pathway that we have found to be abnormal in many autistic children. Many children have low levels of methionine and its product, S-adenosylmethionine, which are essential precursors for cellular methylation reactions. These metabolites also lead to the synthesis of cysteine, the rate-limiting amino acid for glutathione synthesis. We find that the active reduced glutathione levels are decreased and the oxidized form, GSSG, is increased in many autistic children. The only reason that GSSG would be increased in the plasma is that it is being exported from cells under chronic oxidative stress as an attempt to normalize the intracellular redox environment inside the cell.

These metabolic pathways are important not only for redox homeostasis and methylation, but are also essential for error-free DNA synthesis and cell proliferation. So perturbation of these metabolic pathways would negatively affect normal immune function, methylation, redox homeostasis, and clearly affect normal development as well.

We questioned then whether there could be a genetic basis for the metabolic imbalance–increased vulnerability to oxidative stress in autistic children. We are using a targeted approach to autism genetics because we have a phenotype. We are using this metabolic phenotype as a guide to the selection of candidate genes. We plan to evaluate more than 30 genes that affect this pathway to see if some of these genetic variants are increased in autistic individuals compared to controls.

So far we have found an increase in the genetic variant for the reduced folate carrier (RFC1) that regulates folate transport into the cell and would be expected to negatively affect this pathway. Because these genes may affect the synthesis of glutathione, they may be particularly important candidates for gene–environment interactions. We think that the evaluation of multiple polymorphisms that affect a common pathway may provide clues and a plausible explanation for the redox-vulnerable phenotype that we see in autistic children.

A purely genetic approach to autism is a daunting challenge. As you know, it has been estimated up to 100 different genes may be required for the phenotype. Beyond that, there may be different combinations of genes in different autistic individuals. Then if there is a genetic susceptibility that requires an environmental trigger, these same genetic risk factors may be present also in unaffected controls, and that is going to really confound the search for autism-relevant genes. And of course, genetics does not encompass the timing or the severity of the environmental exposures or the heterogeneity that is autism.

Many of us have started looking beyond the brain. Martha Herbert has eloquently suggested this topic. We are looking more at what we call the gut–brain–immune axis. We now know that these three systems do not function in isolation. They are all mutually interdependent; they all talk to each other. We know that all three systems are highly sensitive to oxidative stress, particularly during critical developmental windows.

All three systems are developmentally immature at birth. They require environmental cues to develop normally. So the developmental trajectories of all three systems depend on appropriate signals, and an inappropriate signal could derail normal development.

This implies that you could have a toxic insult to any one of the three systems; it is going to affect the developmental trajectory and the function of the other two. I am trying to think more broadly here.

There are many neurotoxins that are known to equally impact the gut and the immune system. The two that come to mind are mercury and lead. They negatively impact all three systems. A healthy brain needs to develop in the context of a healthy immune system and a healthy gut.

This brings us to new questions. Do we need a broader paradigm for pathogenesis, a more systemic approach beyond the brain, and could there be a component of metabolic encephalopathy that could be treatable? We think that the oxidative stress hypothesis that we are pursuing at least encompasses the possibility of the gut–brain–immune interaction and gene–environment interactions as well.

How do we get from epidemiology to mechanism, which is where we all want to be? We know that the genetic background clearly affects the vulnerability and resistance to environmental insults. We know the environment will alter gene expression.

But we have multiple additive and variable genes as well as multiple and additive and variable environmental factors that make this link to behavior quite tenuous. They are both necessary, but neither is independently sufficient.

So what if we interject a metabolic endophenotype? This can lead us closer to relevant genes. It might lead us closer to and give us clues to relevant environmental factors, and could lead us possibly to a mechanism. In our case it would be redox imbalance or methylation dysregulation that might be more closely related to behavioral abnormalities. Equally important, I think it gives us treatment targets and raises the question, if we can normalize the metabolic imbalance that we see, can we affect behavior?

So for our future research agenda, I would like to suggest that a targeted metabolic signature in fact is an integrated reflection of genes, environment, and nutrition on a relevant pathway of interest. Single metabolites lack the context that we need and global metabolomic approaches are not yet mature. We think that this targeted approach can provide metabolic context as well as insights possibly into molecular mechanisms of the disease, candidate genes, and also treatment targets.

For our recommendations, we would suggest focusing on candidate metabolic pathways to provide clues for environmentally relevant candidate genes, not only in the children—and I think so many of the answers are in the children as well—but we can look also into animal models and apply this to cell models to look for relevant signaling pathways or metabolic pathways, including redox, detox, immunologic, mitochondrial to name a few. We can also use metabolic biomarkers as targets for treatment strategies and treatment efficacy.

I also think we need to invest in the children who already have autism. The treatment options for these children are very limited. We need to invest in placebo-controlled, double-blind studies to try to advance standard of care for these children. Within these trials, I think it is very important not just to look to see if we have a mean difference between groups, but look at which children responded and then characterize those children very carefully genetically and metabolically. Also, equally important are the nonresponders or the negative responders, so that we can begin to individualize treatment for these children because the parents will tell you, each child is an individual case.

I think it is also important to do comparative studies to look for differences in CSF plasma and urine to try to differentiate central from peripheral differences. Quantitative difference in metabolic patterns may be able to distinguish subpopulations within the autism spectrum.

We also need predictive biomarkers. Examples would be to evaluate high-risk children that present with developmental delay because that is usually how the parents bring the children in—with speech or developmental delay. This is a high-risk population as well as siblings and discordant twins. If we can come up with predictive biomarkers that would go along with the behavior—again, they are not going to be specific, but if you have behavior plus a biochemical biomarker, that would be a huge step forward, and would suggest possible targeted intervention strategies.

For infrastructure, we would need many different high-tech analytical instruments and a repository or bank of biologic samples. I would like to point out, this repository or bank of samples is a trickier proposition than it might seem. For our samples, we get them fasting, the same time of day, and they have to be analyzed within a certain amount of time, because a lot of these metabolites are unstable. So the sample preparation, the sample storage, has to be very carefully monitored if we are going to have a sample bank or repository of biological samples from these children.

Thank you very much.

Dr. Levitt: Do we have a few questions of clarification before we open up for the general discussion?

Dr. Noble: Jill, with the comments we have heard from parents today about children who have responded to chelation therapy, have you had any kids that you have actually been able to analyze before and after such therapies to see if these metallic profiles change?

Dr. James: We would love to do that. I am not a physician. There are trials now going on with chelation. I think that would be a natural thing to do, to look to see if they were under more oxidative stress, and when you pull out these different heavy metals, does that improve? That would be a fascinating question, but we are not doing it.

Dr. Newschaffer: I just want to ask real quickly, the model you put up where you showed that the gut–brain–immune systems are all immature at birth and are all developmentally susceptible, are they the only three, or were there others?

Dr. James: The reason we picked those is because many autistic children have gut issues and immune issues.

Dr. Newschaffer: No, I understand, but are there others that meet those criteria? I'm just curious. Other systems, endocrine, cardiac? Are there other systems that meet those criteria, too?

Dr. James: Oh, of course, yes. But we picked those specific to autism.

Dr. Goldstein: I'm just curious, are there data on children right now blinded, where you don't know whether they have autism or not, and you see different patterns? If you took 100 samples of children with autism blind and 100 children who were developmentally typical, could you pick them out blinded?

Dr. James: Actually, that is how I got into this field in the beginning. We were doing a study of children with Down syndrome, and we needed a control. Control children are very difficult to get. So I had this great idea that they could bring in their siblings, because for Down siblings are a fine control. One mother had twins, and one had Down and one had autism. When I looked at that *n* of 1, I couldn't tell which was which. I couldn't differentiate the control. It was very different.

So our very first study was simply a follow-up on that *n* of 1. We now have done over 150 kids, and there is variation. We now do a metabolic screen before we are doing some intervention studies. I would say about 25 percent look fine. But what I look for is that methylation ratio and the glutathione ratio, and cysteine is low in I would say 70 percent of the kids in a larger sample.

Dr. Alexander: Do you have any data showing the stability of this phenotype over time within the same individuals? Second, do you have any more evidence for the specificity vis-à-vis other nonautistic developmental disorders?

Dr. James: Actually we don't know that, but I think when we do it at the same time of day, fasting, over a large number of kids, we see a lot of the same pattern.

In our new study that we are doing, I am going to do two baselines, so we get an idea of the variation, because I don't know that.

DISCUSSION

Dr. Levitt: We have a lot of time for discussion. There are a number of issues that came up. This is a session that focused on environmental factors that impact fundamental biological processes, and many of us get caught up in trying to understand fundamental biological processes.

In the context of trying to understand this, we often don't reflect upon the impact that it has, not just on the individual biological system, whether it is a metabolic pathway or a cell, but on the individual child and the family as well. We talked about this this morning a little bit in the context that there is a lot to be learned and gained from listening to families and parents and listening to clinicians, at least from a basic scientist perspective, in terms of clues.

I think some of that relates to what we were going to be discussing and talking about. One is trying to understand timing of these things that we have discussed. Why is it that there is such neurobiological specificity to this in the kinds of things that we have talked about today, which were general changes in metabolism, for example? Why is timing so important? Is there credibility for this concept of heavy genetic load, and what does that mean in the context of timing and specificity, and how does that relate then to the heterogeneity that we talked about quite a bit this morning, and that we heard about from Sue Swedo? So these are some difficult questions that we are going to need to grapple with.

One of the things that I wanted to pose initially is, sometimes we measure things because they are easy to measure, and we don't measure things that are more difficult to measure. So we make associations and then we assume those are part and parcel of the etiology of whatever we are trying to understand.

So I pose this question to the group, where there was a decided focus on metabolic pathways. There are lots of disorders that involve disruption of metabolic pathways. So while those disruptions and those measures may be valid, where is the selectivity and specificity in the application of this? I want to throw that out initially.

Dr. Noble: I can have a go at synthesizing some of what I heard and some of what we are doing. I think those questions are actually the same question.

When we talk about why is timing important, what most people say is, because something happens at a particular time. That is not a very satisfactory answer, because you want to understand the mechanistic underpinning of why timing is important.

What it appears to us, in the context of what we have been discussing today, is that for progenitor cells, and we are beginning to get data also for differentiated neurons, at times in their development when they are making decisions, am I going to differentiate or not, or am I supposed to have a lot of dendritic growth or not, they become extremely sensitive to these changes in redox state. Their biology is very, very different than other times. They are buffered against redox changes. So it looks to us like that may have a mechanistic contribution.

Dr. Herbert: You laid out a whole bunch of questions. What I am saying is not going to follow what Mark Noble said, but I just want to respond to one of your many points.

In my own brain imaging work, I did a series of studies comparing brain volumes in autism and developmental language disorder, otherwise known as specific language impairment. It was almost impossible for me to tell the difference between those two disorders according to brain volume, except that the autistic brain volume deviations from controls were somewhat more. But it was a question of degree and not a question of kind in just about every single measure that I did.

So this raises some interesting questions that the design of my data gathering didn't allow me to address. There is literature in both autism and in language disorders about the role of immune abnormalities in those disorders. We had no immune data. Immune is related to a number of the other pathophysiological processes that have been discussed such as oxidative stress. We had no data like that.

I personally am not clear that we can answer the kinds of questions posed by the kind of brain data I gathered without having data at other levels of pathophysiology; what is it that makes autism, autism and specific language impairment not so much as autism.

Because there are children who when they lose their diagnosis have specific language impairment, and because I am given anecdotal evidence from some of the baby sibs data that a surprising number of the infants who were at risk for autism who did not develop autism developed language impairment, it makes me feel like there are some underlying relationships going on here whose mechanisms, if we were to explore them—and I believe to explore them we would need to be doing some of the work that we have talked about today—we need to do this sort of work.

I agree also with Jill James about targeted metabolic studies, and Dr. James is much more of a biochemist than I am, and I am not an immunologist, either, but certainly focusing on environmentally sensitive pathways and so forth. If we had a way of giving brain imagers like me and other people who are not metabolically oriented researchers a way of collaborating without a huge activation energy with others who could provide a means to gather these other kind of data, we would have the interdisciplinary capability to answer why it is that something similar at one level is different at other levels. I think that we will need to do that.

Dr. Schwartz: A general comment that I have that I want to pose to the folks who gave these presentations in this last session, which I thought were really fantastic: It seems to me like we got a little bit off course, in the sense that part of the intent of this session was to figure out what we know about the biology of autism and what that tells us about etiology. So I want to pose that question to everyone who thought about it and who presented, as a way of trying to get at the question that we were trying to answer during this session.

Dr. Levitt: So more specifically, is it what do we know about the neurobiology of autism?

Dr. Schwartz: Precisely.

Dr. Levitt: The neurobiology of autism. I want the responders to answer carefully.

Dr. Schwartz: Right, as opposed to the behavioral biology of autism. I am talking about the neurobiology of autism.

Dr. Noble: I can synthesize one part of this. What Martha did not talk about is her extremely interesting studies on myelination abnormalities in autism. Those myelination abnormalities and also what is reported in the literature, abnormalities in the latencies in the auditory brain stem responses, which are also indicative of myelination deficiencies, are precisely as we would predict from our developmental studies and from Jill James's work. They are absolutely precisely as we would predict.

I can talk about that with you in detail, but I have been really struck by how close this concordance is.

One of the implications of that in terms of the auditory brain stem response is that what happens is, you get a spreading of the latencies between peaks. The whole nervous system as we all know is based upon having highly synchronized information transfer. If you have a dismyelination disorder in a nerve trunk, so that you get a spread of information transfer, you have signals being delivered at different times. What is described in the literature for iron deficiency in autism and in a variety of things that we have been studying is that you have the spread of interpeak distances that would be predicted from a lack of enough myelin.

Moreover, we are now seeing as we analyze data in more detail that you have a spreading of each individual peak, which is again exactly what we predict. Some fibers are myelinated normally, others are not.

What happens in the auditory system as it has been explained to me as a consequence of this is that phoneme parsing can become very difficult, which of course ties right into language acquisition.

Dr. Schwartz: So what you are telling me is that neurobiology connects to the physiology, not to the etiology.

Dr. Leshner: Can I just ask, are we moving out of this session a little bit?

Dr. Levitt: No. I think the question was, if we are trying to understand how environmental factors impact fundamental biological processes, it is in the context of what are the fundamental neurobiological or other organ system processes that are disrupted. There is evidence from imaging. I would characterize it differently. I don't think we really know whether it is dismyelination. I think we know there are some long tracks that are smaller and some long tracks that are larger, and we don't know the reason why they are smaller or larger. So that is number one.

Dr. Herbert: I can clarify a little bit. But you finish your summary, and then I would like to say a couple of things.

Dr. Levitt: Do you want to clarify the white-matter part?

Dr. Herbert: Oh, yes, I could do the white-matter part. What my own particular work identified was an enlargement in the white-matter compartment in the outer white matter that we call the radiant white matter, that was most pronounced in the areas that myelinated latest. But using T1 weighted imaging, we had no way of deciding whether this is myelin or anything else that existed in that compartment.

So studies are under way, including my own, of more multimodal imaging to characterize this.

But I want to say that there is some interesting unpublished data, and I have some slides of it that I got from Carlos Pardo. He was inspired by my imaging localization to go back and stain brains in the distribution of gray matter, outer white matter, and deep white matter for the activated neuroglial, astroglial, and microglial cells, and found that there was a greater amount of astroglial activation in the area where I found increased white-matter volume, the outer area, not the inner area, with microglial activation in the cortex, which is not the same thing as saying reduced myelination. This is just a few brains.

So there are a whole bunch of links here which haven't been replicated a lot, but it does raise the question there. You can have from that a lot of exciter toxicity or altered modulation of transmission in synaptic activity, which is another mechanism that would still get you to the same place in terms of disruption of signaling from the very earliest parts of signaling.

So I would agree with your conclusions, but I would argue that there are a number of other ways of getting to that.

Dr. Slotkin: We also have been collaborating with Carlos Pardo with our terbutaline model, and got exactly the same results.

Dr. Herbert: In the mice?

Dr. Slotkin: In the rats, in the same pattern that you guys saw with the autism brains as well, identical.

Dr. Herbert: The astrocytes in the outer white matter?

Dr. Slotkin: Microglial activation in the white matter, cerebellum, and cerebral cortex.

Dr. Levitt: What is the period of sensitivity?

Dr. Slotkin: In the rat postnatal, two through five.

Dr. Levitt: And outside of that they were resistant?

Dr. Slotkin: They were resistant outside of that. It is a window that corresponds to the second trimester of human brain development.

Dr. Levitt: Women don't get terbutaline during—

Dr. Herbert: But they get other exposures.

Dr. Slotkin: ACOG withdrew recognition of maintenance tocolysis with terbutaline in 1995, but it continues to be used. I will quote from a couple of people I spoke to: I use it and it works. Never mind that the placebo control trials show that it doesn't.

But Carlos did exactly the same studies on our rats that you guys did, and got the same results.

Dr. Herbert: But there is also probably a final common pathway in there, just to point out.

Dr. Slotkin: Absolutely, overactivation of beta receptors during that period also causes oxidative stress. So I think what we are seeing, and we have all been alluding to this, what we are seeing is that a whole series of apparently unrelated insults that really differ in the way we think about mechanism in the classic sense nevertheless converge on a common set of final pathways that then produce the consequence. Then it becomes an issue of, can you feed into these common final pathways from fill in the blank, and is the timing right to cause damage to the parts of the brain that are most likely involved.

Dr. Levitt: The other thing that I wanted to say about the neurobiology is that we actually can say what it isn't more than what it is. So in all the studies that looked at changes in cell numbers, for example, or neuronal numbers or changes in cyto architecture, the organization of different parts of the brain, there are no findings that I know of that show even moderate differences.

There are reports of changes in individual neuronal structures here or there, but they are counterbalanced by reports in which there are no differences. But when you look at really careful stereological studies, this is a disorder not unlike other neuropsychiatric disorders, in which there are not profound structural changes in the brains. I think it is fair to say that.

Dr. Insel: There is only one stereological study.

Dr. Levitt: Jeffrey Huchsler has published and David Amoral has published, so there are two.

Dr. Insel: That is the sum total of the literature.

Dr. Levitt: Listen, the cupboard is relatively bare in terms of this.

The other thing to keep in mind that I always emphasize when I talk to students about this is that the neurochemical findings in the brain are all based on measurements that are done in, for the most part, adult brain. For any of us who do basic science, we know that the danger is there of trying to extrapolate a steady-state measure, irrespective of issues about samples or things like that, steady-state measures in the adult and trying to understand what happened developmentally.

The third thing I want to state is that from a neurobiological perspective, studies that have nothing to do with autism talk about development and developmental trajectory, and have quite important findings to tell us now as developmental biologists that we need to pay attention not to what we are measuring at any individual point in time, but it is the trajectory of development, the trajectory of change that really matters.

I think Jake Eades's study from NIMH points to this, in terms of measuring gray-matter volume doesn't correlate with IQ. What correlates with IQ is the change in gray-matter size over time in any particular individual.

Dr. Goldstein: I wonder if you would agree that some of these mechanisms that are being defined are not going to be specific to autism. We are looking at these interesting interactions that several of you have talked about, but I think if you were studying a different neurobehavioral outcome, a different neurodevelopmental disorder, including Carlos Pardo's work, when you start looking at the other nonautism disease controls, they have very similar changes.

So I don't know that these mechanisms are going to get at the specificity of what we call autism. Maybe they are very relevant to aberrant brain development. I'm not sure we yet know why these things could result in the clinical picture of autism.

So as you are studying these mechanisms, and maybe it is a whole nosology issue here of what is autism and what distinguishes it. So we shouldn't lump different autisms together; there are other neurodevelopmental disorders of mental retardation and motor disorders that are going to have very similar pathologies and very similar aberrations in redox and cell proliferation and differentiation, just to confuse this.

Dr. Slotkin: I don't think that is confusing at all. I think it is extremely relevant. Let's posit the possibility that you could have two brains with identical morphological changes, one of which comes from a child with autism and the other of which comes from a child with a different neurobehavioral disability. I would actually find that entirely plausible because the brain doesn't just sit there passively and take a developmental hit. There are things that happened afterward like plasticity and adjustment that are influenced by environment, that enrichment.

We already know, for example, that the incidence of learning disabilities and lowered IQ in the offspring of women who smoke can be completely obviated by an enriched rearing environment. So why should we adopt the idea that we are going to see a morphological phenotype that says autism or neurochemical type that says autism, when the odds are that it is far too complex to be defined within those rigorous bounds?

Dr. Goldstein: I think that is what is happening.

Dr. Akil: I also wanted to say that sometimes, a matter of degree might wind up eventually resulting in a different qualitative difference.

For example, you can have somebody who has a small degree of language problem or hearing problem, parsing problem, but it can be overcome by something, parental training and so on. You can go just one step beyond, so it is still the same problem, just a matter of degree. But it can be so isolating that it can have social implications, emotional implications, family interaction implications, and you could wind up with a somewhat different syndrome. I can imagine that that could happen, where one kind of symptomatology would wind up just as a matter of degree facilitating other types of symptomatology.

So I would be careful about not putting too much stock into this variation at that level, and still thinking about dimensionality. So the language analogy that you are seeing in your imaging stuff I think is very telling, very exciting, and would be very interesting to see in siblings and so on.

Dr. Insel: I think there needs to be a little urgency added into the discussion. I am really concerned that we haven't thought about how to focus this area of science on where the needs would be greatest.

This same forum maybe 6 weeks ago, something like that, with

Dennis Choi chairing the meeting on biomarkers, that was our first such meeting. When biomarkers came up today, some of us are trying to be thinking about that. We were thinking about biomarkers for clinical neuroscience very broadly, and there were lots of discussions about even the difficulty of getting a good biomarker for Huntington's disease, where we have a gene that is actually diagnostic, trying to find physiological transcriptional changes before the onset of symptoms. Walter Korschetz described efforts to do that in a simple Mendelian disorder.

What strikes me here is that the really urgent need for a biomarker that could be as Jill James was saying predictive, something that you could use at 1 week or 1 month or 6 months, well before you have to begin thinking about what the early detection behavioral paradigm might be.

If there is an opportunity for that, I would think that would really drive a lot of the research, and it would be one of the most important things that could come out of trying to identify the pathophysiology.

Dr. James: One caveat to our results is that we are looking at this metabolic profile in children who already have autism, so there is no way to know whether it is a cause or a consequence. We are funded to look very early—see if this metabolic profile is there at 12, 18 months, before diagnosis. This is looking at the developmental delay clinic population, and then following the diagnosis to see whether the ones that had the abnormal profile go on to diagnosis more often than the ones that end up with just developmental delay.

Another important issue is talking about lack of specificity. Do we really care if oxidative stress is an important modulator, whether it be ADHD or what, if we can correct it early, it may impact much wider a population of neurologic disorders rather than just autism.

Dr. Insel: I think we really do care, though. That is one of the issues we talked about. The FDA was at that discussion to talk about how do we qualify a biomarker, something they care a lot about. It was all about specificity and sensitivity. I don't think we are hearing either of those in this discussion. I'm not sure what it would take to get them.

Dr. Akil: But it doesn't have to be a disease, right? It could be a dimension. You could have a biomarker for retarded language development that would work across five disorders. FDA would accept that, just like cognitive problems in schizophrenia.

Dr. Insel: They would accept it for what purpose? For an indication for treatment?

Dr. Akil: Yes, and a biomarker.

Dr. Lipkin: Biomarkers might differ depending on when you do the collection. Given what we know about the blood–brain barrier early in development, you might be able to find things in cord blood that you wouldn't find later. I'm not saying that is the case because we haven't looked. But that is something that we need to bear in mind. So when you talk about biomarkers, it is important to find when we look and how we look.

Dr. Herbert: I think there are a time and place for sensitivity and specificity and a time and place when they are not indicated. I think the FDA is rightfully concerned about biomarkers that are sensitive and specific at times when they are being used to make decisions about interventions that could be deeply harmful to the individual, such as chemotherapy or surgery.

I think that there are other times when the pathophysiology—in this case, it looks like there is a fair amount of nonspecificity to at least some components of it. My own personal hypothesis is that the specificity of the behavioral phenotype may come at the level of network interactions in brain connectivity and not at the level of other things, so that it is a computational outcome more than something that we would measure with the level of biomarkers that we may be talking about here.

On top of that, it may be that the markers that characterize what is specific about autism may not be the same as the markers that characterize where we can treat. I think it is really important to keep that question alive, because otherwise we may be so insistent upon the traditional sensitivity and specificity criteria that we will march way down that path and miss things that are low-hanging fruit, but that are more generic.

Dr. Insel: Dr. Herbert, you should maybe expand on how that might work, because I think that is a very important point for this discussion, just to think through some specifics. What would be some targets, for instance, that could be ripe for interventions?

Dr. Herbert: Let's say you have a child who is 3 months old and starts having ear infections every month. Frequent ear infections are commonly seen in children who later become autistic. Frequent infections set up an inflammatory process and a depletion of redox capacity, which could lead to a lot of different problems.

In my own clinical practice, I take this history on everybody who walks in the door, and I find this history—and this again is not an epidemiological study, it is just my clinical experience with children with autism, with children with ADHD, a lot of that, with a lot of neuropsychiatric conditions, nonspecific.

Is it possible that if you supplemented these children with antioxidants from an early period of time, you would reduce the severity or even prevent whatever catastrophic flipover is involved, if there is such in a metabolic transformation where there is some failure to be able to do whatever it was you could do before.

Some people think of autism regression as a kind of energetic failure. There are various metabolic theories. We don't know enough to say what it is, but if you could address the metabolic depletion that could potentially have been occurring progressively before it went over a threshold, it wouldn't really matter if it was specific to the disorder. The specificity may come from something that is almost incidental to what it is that you can treat. So that is one example.

I think Jill James and some others, Mark Noble, you may be able to comment on that.

Dr. Levitt: Before you get to that, I was going to ask Art Beaudet, have studies been done in the syndrome disorders, independent of trying to link it to an association with the co-occurrence of autism diagnosis in those syndromes, in terms of metabolic studies and other things?

Dr. Beaudet: I don't know. As I listen to this discussion, I think that—

Dr. Levitt: You're not going to answer the question?

Dr. Beaudet: No. I think that there is the syndromic group. I believe they are largely genetic and they greatly complicate trying to study this other group, which is potentially milder, not dysmorphic, and where there is a lot of uncertainty about the etiology.

The dysmorphic group are a lot like mental retardation, and most of them are mentally retarded. So I think if you are trying to ask some of these other questions where intervention is going to be dramatic, it is not going to be dramatic in the individuals who have an underlying genetic abnormality in their neurological function.

So I think those could be weeded out, and then I think these other approaches would best be focused on more normal looking children. I think also, we are to the point where some substantial fraction of the patients can have a pretty convincing underlying diagnosis. If somebody is going to be imaging or this or that or look at oxidation, I think one would want to know, is this patient somebody we know the underlying etiology or not, or is it in the unknown group.

Dr. Levitt: What is the metabolic state with the individual with Rett syndrome?

Dr. Beaudet: My impression would be, looking across all these disorders in mental retardation, that it is extremely heterogeneous.

Dr. Levitt: Within Rett syndrome?

Dr. Beaudet: No, within disorder. But the problem is, probably no one diagnosis accounts for anymore than 1 or 2 percent of the population.

It is like saying a phenylalanine diet works in PKU (phenylketonuria), so why don't we try it in Down syndrome. We understand the pathophysiology in PKU, and you can have some rational input, but I am very skeptical for the group that have definitive heterogeneous defects that there is going to be any intervention that is going to be dramatic other than just supportive intervention, learning processes.

Dr. Swedo: One of the ways we might go after this is, and Sophia Colamarino, you might want to comment on this from your workshop, but just a very simple experiment of taking Fragile X and MECP-2 and other individuals who have autism and who do not, and begin to evaluate the similarities and differences, so your question could be answered about whether there are specific metabolic defects associated with autism by looking at individuals in which we know what the genetic defect is, and then look at the additional factor whether or not they have autism, because not all of them do.

Dr. Beaudet: I think it would be very interesting to particularly compare the children who are at the extremes of some of these diagnoses. But you have to be careful, because in Fragile X you are talking about a male or a female, how large is their expansion. They don't have a pure single genotype.

Dr. Pessah: I think we also need to extend this to those genes or genetic markers that have strong evidence for linkage to autism, the Met gene, Cav1.2, and there are several others that were mentioned today, and construct the animal models and see if they have an inherent oxidative stress when it comes on board, does this influence their metabolic status, and then go back into the kids you have identified with the genetic problems who have the strongest linkage, and who are not as profoundly affected as Rett syndrome, and decide whether or not they are under oxidative stress.

I am very concerned that if you go into every autistic individual and treat with an antioxidant, what about those kids—and I know this study hasn't been replicated—that do better with fever? Talk about oxidative stress as being a cure, or at least a mitigating response.

Dr. Levitt: I wanted to pose another question. Robert Strausberg wasn't able to make it, and he was going to talk about environmental factors and mutations and draw upon his work in cancer, where it is clear that there are certain genes that are more susceptible to the environmental factors than others that cause *de novo* mutations.

I wanted to put you on the spot to deal with this issue a little bit more. I think in the cancer literature, it is very rich in trying to understand the role of environmental factors in perturbing fundamental biological processes. Mostly in cancer it is pretty straightforward, because it is dealing with this issue that Mark Noble talked about, about proliferation of differentiation. But what do we know about that in terms of some of either the candidate regions or copy number variation that has been reported recently, and how environment may play a role in that?

Dr. Beaudet: The thing we know the most definitively to be associated with new mutations is paternal age. That is really dramatic, convincing, and relatively well understood in terms of its molecular basis.

I thought that Dr. Susser gave an interesting example of how, if you were fully deficient and you incorporated U instead of T into your DNA, that this could bring about a risk of mutation rate.

In the case of these copy number variants, there are differences among individuals in the population as to what they have for exact genomic structure. Some of these are prone to having a *de novo* event, more prone than others. There is some of that kind of data evolving, so that there will be slightly unusual rearrangements in a parent that aren't deleterious themselves, maybe an inversion, but then predisposes to *de novo* events in the offspring.

Dr. Levitt: So are there going to be *p53* or *p53*-like genes that we are going to identify here that relate to autism as opposed to cancer? How many different mutations have we identified with *p53*? Hundreds, maybe more, thousands.

Dr. Beaudet: I think there are going to be lots of genes, and they are going to have lots of different mutations in this heterogeneous group. If there is some other residual group we don't understand, then there, I think a lot of these other things are going to—I think one has a better chance of success at testing environmental factors if you weed out the people with frank genetic abnormalities before you start looking at it, or at least look at them separately.

There, I think they have a genetic abnormality that is moderately overwhelming relative to the environmental effect. There will be some other effects, but I think if everybody would effectively weed out two ways—with lab data it is somewhat possible, although the lab tools are still pretty primitive, and on the basis of dysmorphic features.

Personally, if I wanted to look for an environmental effect that would affect the possible impact, I would want to deal with male patients who looked perfectly normal and go in that population. I think we would have a much better chance of finding something that is making a difference in that group.

Dr. Noble: I think there are some conceptual issues there that are quite strikingly important. What we need to ask is important, but the fact is, because the FDA wants it doesn't mean it is the right question, with all due respect to the FDA.

For example, to take a cancer example, *BRCA-1* is a great predictor of whether or not you are going to get breast cancer, but only for those patients who have a mutation in *BRCA-1*. It identifies a small percentage of the individuals who are going to get breast cancer. It is useful to have as a diagnostic and may put you on different paths. You can get to the same endpoint in different ways.

Let's take *C-Met* as an example we discussed. In an individual who has compromised *C-Met* function because of a mutation in it, from the little understanding we have now, from what Pat has published, and we have published some other stuff in the literature, would we be able to correct that with some kind of an antioxidant therapy? I don't think so. If the compromised *C-Met* function is because of what Jill James is saying, then we may have a real shot at having a correction.

So I like this idea of trying to understand how to screen out the genetic populations so that we can focus attention on the other aspect. But I think we have to understand, particularly from the stuff that Jill James has presented, that a lot of the genetic mutations may actually be giving us these metabolic disorders that even though they are genetic are still going to be broadly treatable by metabolic modifications. So maybe the FDA is going to ask us for something that for here isn't exactly the right question.

Dr. Susser: Can I say one thing about environmental? Just to extend what Art Beaudet just said, when you separate the group, and I do think there will be a reasonably sized group that will have genetic mutations, they will have heterogeneous genetic mutations that will have large effects, even if they are not completely deterministic of the disease.

But that is a particularly interesting group in which to look for environmental antecedents of those mutations. That was implied in what I was saying. Studies of environmental factors are particularly important in that very group.

Dr. Spence: One of the things that we talked about in our session and you brought up was the idea that we have to understand whether those mutations are in fact functional. So the determination of these variants, because there is a lot of evidence, if you look hard enough you can find lots and lots of different polymorphisms. The question is what is the relevance?

Just a case in point, Fragile X permutation status. There was a great paper that Randi Hagerman's group did that said this is associated with autism, even in the permutation. She can show, because she looked at the RNA, that those kids with autism with a permutation have differential RNAs.

But on the other hand, it turns out Fragile X permutation status is actually very, very common. If autism is 1 among 50 and Fragile X permutation status is pretty common, it could have just been the two are unrelated.

So I think we have to be careful that one of your gaps was assigning functional function to these polymorphisms, and I think we can't do that.

Dr. Noble: That is exactly right. We know so little about this area. From a historical scientific viewpoint it is fascinating, because it looks like early entry points to a number of fields, except there are all these other data that one can talk about, that you think you should be able to understand. It is a bit overwhelming.

Something I am concerned about is that I am really listening to the urgency from the patient representatives who we have here. As we have experienced in all scientific fields, we are always trying to balance our step-by-step progression in science with that urgency. I do have a concern that we have to be running both tracks at the same time.

It is too early to say that this is right.

Dr. Leshner: I've been waiting for somebody to say that, that is, to take either exclusively a general biomarker or a specific biomarker approach seems very dangerous to me, particularly given that we do eventually have to get to the specificity of the specific disorder that we are talking about.

Dr. Insel: I think there is a crosscurrent here that still needs to be clarified. If I am hearing this right, we are getting two messages today. One of them is the general issues that have to do with metabolic stress through development, which is almost certainly not specific to autism, but may be a robust finding that could be a signal for an intervention of some sort, but it doesn't tell us all that much perhaps about the specific pathophysiology of this illness.

The other current that we heard more this morning was that autism itself is so general and it is so many things that even there we need to drill down and get much more specific, much more selective. We talked about doing *n* of 1 studies using individuals as their own control, defining kids perhaps by what they responded to and calling that a subgroup.

I do think we have to get clear about where the most traction will be going forward, because those are two very different approaches, and we don't have enough money to do everything.

My own bias is that if you look at other areas of medicine, generally you see progress best. Asthma is such a great example. When you can find a way to define a subgroup even beyond what you see clinically, and come up with groups that can be defined by some pathophysiological variable that now allows you to go after these other factors that help to grow that out, I don't think we are going to get there if we start taking very general kinds of pathophysiological markers that don't in any way take us to this endpoint.

So just an opinion, but based mostly on what I see in the rest of medicine.

Dr. Beaudet: I'd just like to comment. I'm sure I come at this from a very genetic perspective, but to me the most urgent question, which it is obvious has been urgent for more than a decade, is whether the incidence is really changing or not. This is not rocket science. I think it is the CDC's area.

If the incidence is really changing, there are environmental factors, there are things to be prevented, there are interventions to be done. If the incidence is not really changing, that is a very different situation.

I think also, my impression is if the incidence is changing, the percentage of males relative to females should be rising with that, because that is the group that is going to be more likely to be involved in environmental interactions. But I think the most urgent question is to know if the incidence is changing or not.

Session IV
New Approaches and Discussion with Workshop Attendees

Participant: My name is Kelli Ann Davis. First of all, I want to show you my son, Miles. He is now almost 15 years old. When he was a baby, I think autism rates were 1 in 5,000, and they are now 1 in 150. This is my son at two and a half months old. If you look closely, he is trying to mimic speech. Two and a half months old, he was completely fine.

Here he is at 1, and he is completely fine. Here he is at about 3 years old, and if you look in his eyes, you can see he is not really there anymore. Here he is holding his baby sister when he is 6 years old, and you look at him and look how sad he is.

My son is now almost 15 years old. There are a lot of smart people in this room. I'm just a mom, but I am asking for your help to find out the truth about what happened to my son. I believe it had to do with vaccines. I believe mercury had something to do with what happened to my son.

I am here for the truth. That is what I have always wanted to pursue, is the truth. I am just encouraging you all to remember when you are talking, it is about our kids. I have got to tell you that the first time I heard Martha's talk a couple of years ago at the symposium, I had to go upstairs to the hotel room. I couldn't even hear her talk because I thought about what has happened to my son and what was going on in his brain.

You are all scientists and you are looking at it from the scientific perspective, and we need that. But there are thousands of parents out there who are heartbroken, and when they hear the descriptions of the brain and the white matter, it is almost too much to take.

So I guess I am just pleading with you all, first of all I want to thank you all for being here, but to please keep that in mind, and remember the kids who aren't babies anymore. We need the help as quickly as possible. I just appreciate everybody being here. Thank you so much.

Dr. Leshner: Thank you for that. It underscores the urgency I think we all feel. So thank you.

Mr. Blaxill: It also suggests a tiering around what kind of biomarkers we really care about. They are diagnostic biomarkers that are final common pathways, they are prognostic biomarkers, they are treatment response biomarkers.

It is easy to have prospective studies and think about the kids that aren't diagnosed yet or are not affected. The constituency out there needs attention to treatment response. It may not be now, and it may not be specific. It may overlap with all sorts of other things, GI diseases and the language impairment, but there is a prioritization implied in the biomarkers.

Ms. Redwood: I just wanted to make a comment, too. I guess this is moving into the general session, since the parents are talking. What I have heard today is that we are looking at the potential for there being an environmental toxicant that may have caused our children's disability. One of the questions I have for the panel is, why aren't we testing the children? We went into Brick Township and we tested the water and we tested the dirt. We tested everything we could think of, but nobody ever tested the children.

I hear over and over again that mercury is one of those metals that might be causing this. I know for my son, he had over five times EPA's action level of mercury in his body. I am just wondering why we are not testing the children. If there are multiple toxicants, let's look at the kids and see what they have. If it is mercury, lead, PCBs, to me that is the study that we are ignoring right now.

So I would like to ask the panel if anybody is looking at that, if anybody is doing urinary porphyrin levels in these children, what are the plans to test the kids?

Dr. Swedo: I have mentioned that the NIMH M.I.N.D. phenome project is the pilot, and that is absolutely child focused, family second, home third. My colleagues can talk about their own studies, but I think your point is very well taken that if you are going to find it, you need to look where they are affected.

That is one of the tensions between the need as Art Beaudet talked about, to find out what the change in incidence is. Those are large-scale expensive studies that have to be done. On the other hand, if this was leukemia or another medical illness where we could look at a cell system and know exactly what was wrong, we wouldn't spend a lot of time looking at the unaffected to figure out what was happening with those subtypes.

So I think if I were speaking to the urgency, and I am trying to, my plea would be that we do this kind of meeting where everybody is using the same platform where possible, and drawing on the strengths of each of the different kinds of advances.

Dr. Schwartz: I agree with what Lyn Redwood just said. I would go further by saying that what we should do, we should try to figure out what studies are underway in populations of kids with autism that we can build in the state-of-the-art, but admittedly somewhat limited environmental measures that could be done on the biospecimens that are available within those studies. There is no reason we shouldn't do that. If that takes expanding the studies, we just need to look at what it will cost and try to pull those funds together to use in the best way possible to make that information available.

The question I was trying to get at this morning is, what are the cohorts that are ongoing that we could leverage to append these additional studies to that would get at the answers? I think they are not perfect. As Dr. Insel pointed out, we don't have 39,000 assays, we have 20 assays right now, but maybe 2 or 3 years from now we will have a thousand assays. While we have 200 assays, we may as well make use of them.

Dr. Falk: This has been a very interesting progression from Tom Insel's comment all the way through to here. It seems to me that a large part of today, the discussion has been around mechanisms and pathways which could be impacted very significantly by environmental agents, but not nearly as much discussion about specific environmental agents. And of course, the pathways themselves may not be fully specific.

In truth, there have been various times where environmental etiologic factors have been identified even before pathways are understood, and only afterward does one go back and understand the pathway.

I guess one conclusion that I do draw from this is that perhaps there ought to be certainly more attention to specific environmental factors, both experimentally as well as in terms of—in the epidemiology we will discuss tomorrow morning, we will have the opportunity to see just what those opportunities are.

But it strikes me that that is an important area that, as I am seeing this all put together, is not fully addressed, perhaps.

Participant: I have a question for Dr. Noble. You talked about thimerosal toxicity at the very end of your talk. Are all of the oligodendrocyte precursors selectively vulnerable, more so than other types of brain cells?

Dr. Noble: This gets to the issue of when in the life period of a cell or lineage you see vulnerability. If you wanted to design a system to enable you to study problems like this, you would come up with something like what we now know about the oligodendrocyte lineage, because it has so many advantages to doing this kind of work.

One of them is that myelination occurs at different time periods in different parts of the nervous system. The organism creates cells that intrinsically have different timings. One of the ways that we learn a lot about these problems is to try and understand what controls those timings, and it has turned out to be this intrinsic redox biology.

So when we look at other cell types like embryonic cortical neurons, we find that similar principles apply, but we have to study them at the right stage in order to do this.

I think in terms of this issue of specific toxicants, although I think this is an oversimplification, I have to say that what we keep seeing is that all the cell cares about is, is it oxidized? It doesn't care who is doing the oxidation. Obviously at other levels there are chemical specificities, but this is what we are seeing.

Can I ask a question? I am trying so hard to understand this area. I think I see an experiment, but I want to ask whether it is a good one, that ties together some of the things that we have heard.

Following on from Dr. Schwartz's and Ms. Redwood's important comments about looking at the kids, and what we have heard from the parents about those who have used chelation therapy, is it a good question to ask, if you have a child with autism, and you now screen these other parameters, mercury load, lead load, PCB load, get an environmental toxicant profile on them. Now you do the chelation therapy. Is it the case that the kids in whom that works, are those kids in whom we have higher levels of heavy metals? And is it the case that the parameters that Jill James's studies normalize, or that auditory brain stem response normalizes? Is that a type of focused question that one can ask to get some traction?

Ms. Bono: That is basically the recovered kids study, which you are talking about. There are DAN doctors throughout America who have kids whom they started working with, 3, 4, 5, 6 years of age, specifically the younger ones are the ones that have the best recovery rate. Some have Jill James's profiles, these kids have shown methylation problems, oxidative stress. They start chelating, but they also do other things, giving them glutathione, cysteine, all of the things to help with that.

So there are those entry-level treatment biomarkers that the doctors have when they walk in the office. Then they have it tracking as they go along.

Dr. Noble: Can that data be made available to us?

Ms. Bono: The DAN doctors have said that they would be very willing to have data mining go into their offices and pull that type of information.

Ms. Redwood: There is one of the clinicians I saw here a few minutes ago, Nancy O'Hara, who has a very large practice, who might want to share her data.

Participant: I am Nancy O'Hara. I have been working with children with autism for 25 years, first as a teacher, the last 9 years solely with children with autism.

First I want to thank the researchers, because they have given us the information that we need to see why our kids are biochemically sick, and they are, but also to see how we may begin to treat them. They are treatable.

We do have that data. We have the urines, we have the stools, we have the leads pretreatment and posttreatment, in recovered kids and in kids who are not recovered. Believe me, it is not 100 percent, but the data are there. We need help mining that data and taking that data from a large group of clinicians now who have it, but we just don't have the resources to then pool the data together and use them. But we have them.

There was a lot of talk this morning about inflammation and also this afternoon. They mentioned tonsils on one of the slides.

David Gozal from Kentucky has very interesting literature on the very damaging effect of repetitive hypoxia. He has looked at it in cell culture systems and also in clinical systems, and found that children with learning disorders and also sleep disorders often had tonsil problems, and when they were removed they actually improved considerably. So his repetitive hypoxia paradigm would also fit in with some of the redox type of studies people have been talking about, and I think should be thrown into the big picture.

Dr. Swedo: Before we go down that field though, as a pediatrician I just have to remind folks that tonsils in a 7- to 8-year-old are very different than those in an infant and neonate.

Dr. Leshner: Somebody here was going to respond to the last question.

Dr. Akil: It was a comment about physicians and other people in the community who have information. Tom Insel and I were talking at lunch about how one might engage physicians in the community, whether it is part of a CME (Continuing Medical Education) or volunteer or whatever, whether we need a medical informatics national project that sits back and thinks about what kind of information is needed to do this in a systematic way, meaning something that one can participate in where the kinds of information that are needed, the kind of diagnostic criteria that are required, the kinds of treatment, the kinds of levels, what assays are approved or not approved, whether we can put something like that together either in a trial in a few centers to begin with before expanding it, and bring information into it and see if there is any way to begin to rely on people who are doing the footwork but feel isolated, and have scientists who are good at data mining or analyzing.

But if you patch it like a patchwork, very pell-mell, it would not be useful. We need to come back to you and say here is the information we need, and then get it from the people in the trenches.

Participant: I agree with you, but you also have to address the urgency which a lot of these parents feel. If you start prospectively and ignore all the data that are already there.

Dr. Akil: You eat it. You eat what you have. You eat the data that you have so far.

Participant: But you have to use that, and they may not be as clean as the data you want to use prospectively. But I think you have to use some of the data you have now to be able to start.

Dr. Swedo: I think that is a fabulous idea. We thought it was such a great idea that we started 3 years ago to develop a national database for autism research, which allows clinicians in the field to become researchers by providing them with the clinical tools they need to do systematic assessment of their patients.

Our group has been very impressed with the DAN practitioners and are grateful to them for what we are learning from them, are hoping to partner with them even more in the future. But in addition there is another network, the Autism Treatment Network (ATN), which started out on the West Coast and Boston. They now have a dozen sites. They are hoping to get 20 different sites, both academic and clinic community based. They are gathering data from their patients, and if Paul Law is still here, he can speak to the Autism Speaks registry, IAN, Interactive Autism Network, that allows parents to input their data directly about their kids and get instant feedback.

I think we are poised from an informatic standpoint to meet this need of urgency and get the data very quickly, start looking for similarities and differences across this group, and then do more in-depth systematic study as the patterns emerge.

Ms. Bono: I agree with you, all three of those things are very good. With the DAN doctors, with the huge practices, they need to have more of a systematic approach, where three or four people are on the payroll, and they come in and they mine that data based on whatever criteria they have, and then they move to the next one. These doctors just don't have the time to go back and try and put it together and give it to you, but it is there.

Dr. Insel: I think one of the great things about the National Database for Autism Research is, it does give you the standardized assessments. All the tools are there in place, and anyone can use them anywhere. It is totally public access, or will be in September when it goes fully live.

The relevance to this meeting specifically goes back to David Schwartz's and Lyn Redwood's point, though. What we don't have are the large repositories of biological samples on all of the thousands and thousands of kids who have been treated. They may exist someplace. We have a relatively small brain bank, we have small banks of other kinds of samples, but clearly there is a need to do here what was done for childhood leukemias 25 years ago. You find a way to organize, consolidate, and then go from *n*'s of 10 and 20 and 30 to 2,000, 3,000 or 20,000 or 30,000.

In a complicated area like this, you are going to need those kinds of large *n*'s to be able to find the subgroups that really will give you ultimately that rigor. We will have the clinical piece. One of the things that would be great for this group to weigh in on is what would be the biological samples, when should they be collected, and what would you want to do with it.

Dr. Leshner: I think that one we should hold for tomorrow.

Dr. Schwartz: And maybe what could we do with them in the absence of an absolutely ideal study. We have probably at NIEHS two or three epidemiology studies that we are funding in autism. I'm sure you probably have a half dozen or a dozen or in aggregate.

There are a number of epidemiological studies in autism that have been done. They are not using the same tools necessarily, they may not have the same diagnostic criteria, but there are areas of overlap that we would agree are critical elements across all of those studies that could serve to bring those studies together in a biobank that we could then mine for environmental data and genetic data and other data that could push the field forward before the ideal population has been acquired.

Dr. Falk: I am very supportive of these ideas that are coming forward, particularly if the chelation data, for example, are one of the strongest indicators for there being environmental agents. They should be looked at in detail for any group like that that is thinking about environmental agents.

But if I may go back to something Mark said before. I want to make sure I understand this correctly. I understood what you were saying to be that there are so many environmental factors which could conceivably affect redox status and pathways, it is almost immaterial to look for the environmental agents? I don't know whether you quite said that, but maybe that is what I was thinking. You were implying I think that the specific environmental agents might be so numerous that. . . .

Dr. Noble: No, it is the second one. I think that if one wants to test the hypothesis that mercury is the primary causative agent in autism, that that is the wrong hypothesis. If one wants to test the hypothesis that mercury is one of many environmental factors that may contribute to this outcome, that looks like the right hypothesis.

So if we look at what we can look at now, there is a limited number of agents where the sensitivity of analyses are sufficient to enable us to do reasonable studies, the organic materials, PCBs, a few other effects. That data may turn out to be extraordinarily compelling, particularly because of what we are hearing about the heavy metals and the chelation therapy. At least heavy metals are something that can be analyzed pretty well.

What I am specifically concerned with is that—with all respect and admiration and concern for the parents' groups and everyone who has been trying to pursue the idea of a specific environment toxicant or a specific vaccination, just from a biological point of view, it doesn't sound like a great hypothesis. It sounds like these may be pieces, that they happened at a particular time, but they are not going to apply to all the kids.

From the cell's point of view, I don't think it matters. Am I oxidized because of an inflammation? Am I oxidized because I got mercury? I don't care, I'm oxidized, I am in electron deficit. The data that I am hearing just keeps agreeing with that. Even this idea of astrogliosis and the white-matter tracks, when we take these oligodendrocyte progenitors and expose them to oxidative stress, they turn into astrocytes. So even that is a really intriguing outcome. We have to look at specific astrocyte populations there.

So that is what I mean, that these all could be players. It doesn't sound likely at this stage that any one of them is a player of central importance.

Participant: I am Lee Grossman. I am president and CEO of the Autism Society of America, and more importantly, the dad of a young man, a 19-year-old with autism.

First an observation. I want to thank all of you for this wonderful assembly. Some of the best minds in the scientific community are working on this. The autism community, the parents, are very grateful for your efforts and everything that you are doing.

I think there is one oversight here that I do want to point out. For future planning purposes, I think it is essential for a person on the spectrum to be included on the panel as well as the planning. Hopefully that will be corrected as we move forward.

The comments I am making now have been supported by some of the other people here. I have been wanting to make them since Sarah Spence presented her observations this morning, when she presented the information in terms of the variations in autism, which could be in terms of millions, perhaps billions, of variations out there, when you include all the extraneous information that is out there for environmental interventions as well as the genetic components of this.

Then Tom Insel presented his two models exploring this, and bringing in the phenotype data as well as looking at it, which would be certainly meeting the scientific rigor that I think all of you want to meet.

I wanted to propose—and I think some of the people here, Nancy and others, have started to discuss this—a third model that I think would fit well into that, and that is a treatment model. What you are talking about here is wonderful and it is what needs to be done obviously, but we are looking at another generation of children as you do this. I don't think the community can wait any longer. There is enough anecdotal and proven information out there in terms of treatment that should be explored and followed and implemented.

I think if we put into the two models that Tom recommended this morning, and also incorporate a treatment-guided model into that, the three can work in collaboration with each other. We can develop treatment protocols that can, I believe, meet the scientific rigor that you are looking for, where these kids can be evaluated, we can see what is working, what is not working, and then move forward in that regard.

In the meantime, kids are getting helped, they are getting improved, and we are learning. I think we are going to accelerate the pace of our knowledge and our learning, and certainly help the folks that are out there today.

Thank you.

Mr. Blaxill: I just want to amplify that, and just underscore some of the scientific deficiencies of approaching that. A lot of what we like to call evidence-based medicine is designed to ration the access to market of small molecules that pharmaceutical companies sell. That is a very useful rationing procedure. It is good for safety management and that sort of thing.

I think the types of therapies that we are talking about here with environmental illness are more regimens and ways of life, diets, things that are less invasive and less dangerous than some of these things are, potentially dangerous. So there are special methodological problems, and I think it just argues for some degree of risk taking, comfort with complexity or messiness.

I hate to argue for relaxing standards, because I don't think any of us want less rigorous work, but we also need to be roughly right rather than precisely accurate 20 years from now. I just think it is important to come to grips with the special types of therapy and regimen interventions that we are talking about, so that we don't throw out the baby with the bath water. I can imagine all sorts of negative studies coming out that miss the main point. Sue Swedo, I have talked with you about this.

I was saying to Pat Levitt, some of us, because we don't know what else to do, we can't wait for the clinical science to take 20 years to solve all these problems. We have to act today, we have to act on some model. We don't know whether we are right, but my daughter is 11; I can't wait that much longer.

I was saying to Dr. Levitt, it is like advertising. I'm sure that 50 percent of the therapies that we are trying are absolutely worthless. I just don't know which 50 percent, and sorting that out is a challenge.

I just call that out as a methodological challenge, a scientific challenge, because there is a real risk that you get a collision between the request for rigor and the movement and all of this enthusiasm about helping kids. Those ought to come together and be mutually supportive. It is the potential for them to get antagonistic, and that is something to keep in mind.

Dr. Leshner: I think you articulate well the obligation that the scientific community frankly has to help you meet that need. I like to think, I hope to think, that that is what we are doing here. I hope that this will in fact significantly move forward the research agenda in some way.

But I think the point is extremely well put, and it underscores the obligation that I think everybody feels. But it is good to articulate. Thank you.

Dr. Levitt: It is also underscored by—when you look around the table at the scientists who are here, how many scientists here who are actively doing research started doing research or were trained to do research in autism? Raise your hands. Two. That is a reflection that there has been a sea change culturally in the way science is getting done in a lot of disciplines, that this is a poster child for the willingness of scientists to not do the kinds of standard things they do, which is to keep looking for more and more rigor and being unwilling to take some stands and work together, and come from different fields.

So I think that is happening pretty rapidly, and it needs to happen more obviously. But I think that is a reflection of what you just said.

Dr. Noble: I think we are trying to find out how to meet you, if not halfway, some way in the middle. There are 14 million kids in the United States with some kind of neurological disorder, and parents are trying everything. From the point of view of someone in stem cell medicine, I am desperately interested in keeping this stuff regulated, because there are so many cowboy clinics out there.

But most of you are not doing stem cell transplantation, you are doing things that have less of a risk. You are going to follow multiple areas of research. But if I look at this as a scientist and I can look at a minimum dataset that says, here is a kid, here is a metabolic profile, here is a heavy metal profile, here is a toxicant profile, here is what their behavior is like, and you do whatever you do, and we get those measurements at the end and can say, the kid is behaviorally changed, are there any of these measurements that have changed? We may learn quite a bit from that. It would be nicer if there were standard protocols that people were using.

Dr. Herbert: I really agree with what Mark just said. I think that it is unrealistic at this stage to try and discipline people into specific standardized protocols, since we don't know how to characterize the heterogeneity, and because people are going to do what they are going to do, anyway.

I would propose from a sociological point of view that there could be usefully some support for the self-organization of the parents and some of the treating physicians, like the DAN docs, like Nancy and others, to have more support to build a platform of communicating what classes of data are available, a status report of what is being collected.

It is not enough for you to stand up and say for 30 seconds that you have these kind of data. I think it would be really nice if we could have support for a description in more detail of what is going on, focusing in classes of data to facilitate the interface with NDAR and other kinds of data collection mechanisms.

I'm not clear that it would work coming from NDAR to the parent and treating community. I think there needs to be some support for the treating community, which is exhausted and overworked beyond all description with this incredible burden of cases. It is not just people coming in and mining data. There needs to be preparation for that, so that there is some kind of a systematic approach.

So I think to make this happen, in order to meet in the middle, there can be a transitional infrastructure of setting up what it would mean to do that. Otherwise the activation energy to make it start happening isn't going to happen.

Dr. Beaudet: It seems to me that it would be interesting to know if the two camps could come together around a truly blind chelation trial, in which certain patients got infusions of placebo, and this went on for a year. Some parents would have to take the risk that their child might or might not be on placebo chelation for a year. But I think this would take considerable investment of both sides to agree to something like that. I would be pretty impressed if such a study could show something is going on.

Dr. Swedo: I just wanted to say that such a trial has been developed in collaboration with the DAN practitioners. We are using their protocol and breaking those elements down. The gluten-free, casein-free diet is already under investigation at one of the START centers as well.

So I think the individual components, the hard question and the thing we really have to grapple with is this issue of—it appears that one of the successes of the DAN approach is that it is very broad and deep. There are a lot of things going on with those kids all at the same time, so trying to figure out which components of it are useful is something that is going to take some more work.

Dr. Oberdorfer: Just to follow up on Martha Herbert's point, when you have observational studies that you are going to be undertaking that you haven't planned yet, if you do what you suggest, you are going to see what sort of samples that you can take from a number of different studies in a much more global sense. That way you would have some commonality in a tool kit.

Right now, my impression is that they are going in cross purposes. They are not collected in the same ways, they are not in the same times, but they are moving in that direction. It seems to me that even if the studies go in different directions, you will have samples that you can compare homogeneously. I think that is very important, these kinds of toolboxes. We can do that now.

Dr. Lipkin: Alan, this is not a comment. Maybe if Sue Swedo could summarize for us what is going on, we might save some questions. We are continually going back to asking what is being done at NIMH. If you could just summarize what is being done in terms of treatment, then maybe some of these questions would already be addressed.

Dr. Swedo: All right. The new intramural research program is about a year old. We started with two major types of studies. One is an in-depth phenotyping effort, making use of the anecdotal literature and the clinical experience from clinicians across the country, but also the CHARGE, CADDRE, and other data that had been collected. It is everything from family history of medical illness and environmental exposures to neuroimaging, genetics, and other evaluations.

Within that, we have a regressions substudy that looks at children specifically with regression for additional factors, such as microbial triggers or inflammatory responses. We also have a treatment component. Intramural does best novel treatments. Tom Insel called us a SWAT team. We go where we see a lead. For example, we are using menocycline for its effects on NF kappa B to try and see if that would decrease neuro inflammation.

We have a trial in antiglutamate agents, seeing what effect that would have not only in repetitive behaviors, but overall autistic behaviors. The chelation study is currently on hold because of some recent reports of a rat study that reported cognitive deficits in DMSA-alone treated animals. We are going back to the IRB (Institutional Review Board) on May 1 to look at that question.

That is what we are doing in-house. My colleagues from extramural can talk about the new A centers. But I think that many of the things that are happening, some of them are underway. Probably the most important is this issue of common measures and trying to get as much richness of clinical data as we can from every subject that is studied with NIH funding. That includes, as I mentioned, standardized diagnostic assessments, behavioral assessments, neuroimaging if it is being done on a common platform, as well as obtaining genetic material and biological samples.

Dr. Lipkin: A constellation of toolboxes.

Dr. Swedo: Right. I think we have already heard about some of the ongoing efforts in which supplements are being done to get the same kind of biological data. Now one of the questions is how do we organize it and go after the hypotheses.

Participant: My name is Becky Peters. I have worked in the autism community for the last 5 years. I missed the very beginning, but I don't think until Sue Swedo just mentioned it that I have heard anything about the possibility of food allergies.

I read a lot about and heard lectures on things like the gluten- and casein-free diet, the specific carbohydrate diet, and how for some children with autism, it has not only improved some, but even caused recovery in some.

So I was just wondering if anyone in the research community is looking into the possibility that food allergies or certain foods that maybe children are genetically predisposed to be more sensitive to could be environmental factors in causing autism.

Dr. Leshner: Does somebody have a very brief answer?

Participant: A brief answer is the recent data on microflora associated with obesity, for instance. It is something that occurs in response to a specific diet. We are finding that there is increasing research that tells us that you can change metabolism and adipose cytokines and adipose tissue can change in response to the diet in conjunction with the microflora.

We have the opportunity now with the tools that exist to begin to explore those types of issues. It may not be the standard allergies. It could also be other models that we need to also think about.

Participant: Claudia Miller from U.T. (University of Texas) Health Sciences Center in San Antonio. We have worked extensively with adults with food intolerances and environmental intolerances. Until you eliminate all of the things that bother that particular person, you don't see the problems reversed.

Now we have started doing the same things in autism. The caution is that if you start doing a few things and just gluten and just casein, you may get some reversal but you may not get all of it. You may get other intolerances that develop, which is why you have to have a very comorbidity protocol.

Participant: My question is about it being part of the etiology. I know that diet is out there and it is helping people, but I don't feel like anybody here has addressed the possibility of the food allergies causing that problem. I was just wondering if anyone is considering that, or if that is not on the table for research.

Participant: I am Dr. Richard Deth from Northeastern University in Boston. There is something missing here from this discussion that unifies many of the observations and the questions that have come up during this afternoon's discussion.

That is reflecting some of the work that we did and we continue to do on the dopamine D4 receptor. The D4 dopamine receptor is involved in methylating membranes of neurons. It uses methyl groups from the folate pathway through the enzyme methionine synthase. We discovered that a certain number of years ago.

The D4 dopamine receptor is linked to ADHD, and is now recognized as the most important genetic risk factor for ADHD. The D4 dopamine receptor is linked to lead toxicity and the role of lead in contributing to ADHD. The D4 receptor is linked to IQ. It is a risk factor for IQ reductions. So it has all the characteristics of a candidate receptor, dopamine included.

This is the only receptor that utilizes sulfur pathways. It uses the enzyme methionine synthase that is turned off by oxidative stress. When oxidative stress occurs, be it mercury or be it pollutants or be it pesticides, that enzyme turns off to make more glutathione and robs that system of its methylation ability.

The role of the D4 receptor is to synchronize that gamma synchrony that is gamma-frequency synchronized, synchronization of the brain during attention, a system that is deficient during autism as well as ADHD.

As we have pursued this line of investigation, we have recently found that there is alternative splicing of the gene from the mRNA from methionine synthase in the human cortex. We have found that this alternative splicing is related to aging; it is complete in 80-year-olds, and it is incomplete in 20-year-olds.

As a result of recognizing the central role of this process, I think you can find mechanisms to explore. The enzyme methionine synthase utilizes methyl B_{12}, which is a treatment that approximately 30 percent of people respond to.

So I would say that there are all the elements here to start building from, even though it might not be a complete story. I just recommend that area of science to the panel members, because it can unify many of the things that they are concerned about.

Dr. Leshner: Have you published it?

Participant: I published several papers. The first paper about the D4 receptor was published in *Molecular Psychiatry*. The second paper was about methionine synthase being inhibited by ethanol, mercury, lead, thimerosal, because glutathione levels are low and methyl B_{12} synthesis is impaired. There is a lot to know about this. You just have to look into it.

Dr. Leshner: Thank you. We will put that onto the list. Very helpful.

Participant: I am from CDC. I want to make three points and try to make them quickly. The first is, I hope that everyone will be here tomorrow morning, because there will be several epidemiologic studies that will be presented, and they can be built upon in terms of specific environmental questions that are not being addressed. The CHARGE study will be presented and the CDC CADDRE study will be presented. It sounds like a lot of the questions that people are asking about studies and cohorts might be answered tomorrow morning.

The second point is, Dr. Schwartz was asking about large cohorts that are available for study. I wanted to mention one in China. These are children of mothers that received folic acid. These children are about 12 years old now. There are about 200,000 children that are going to be characterized for autism. So I just wanted to go on the record to say that is a cohort that could be studied.

The third point is to talk about the National Children's Study a little bit more. Dr. Landrigan did mention that study, but there are a lot of environmental agents that are going to be studied as part of that study in terms of levels in the children, and autism is an outcome.

There will be a research protocol that will be available for public comment. If that study is not addressing some of the questions that people have, if we don't have the right chemicals identified, if we don't have the right confounders, mediators, and modifiers described in the study, then I would encourage people who are here today to comment upon that. That study was designed to answer a lot of these questions about specific environmental exposures.

So I just wanted to remind people that there are some studies in addition to the ones that Sue Swedo had mentioned within NIH. There are some studies underway that might be able to shed some light on some of these questions.

Dr. Leshner: I'll just reiterate your point about tomorrow morning. A lot of very important talks are going to touch on an array of these issues that we have been talking about already. But we allow free talking. Thank you.

Participant: My name is Harold Grahams. I am a private practitioner in Pennsylvania.

To address Dr. Insel's issue about the urgency of a biomarker, there is a tool that has come across my radar a few years ago that has been used in chemistry labs and hospitals and university settings all across the world that has been underutilized in autism. It is a high-performance liquid chromatography. There is a gentleman, Wayne Madsen, from PSA Labs who did work with lead studies 20, 30 years ago. Wayne Madsen did some unpublished studies with a controversial group up in the Philadelphia area probably about 10 years ago. I think it is a tool that might answer a lot of questions that all of us as practitioners and anybody could use as a reliable biomarker.

What Wayne Madsen found with a group of kids is that we could give him bloodwork, and he could tell us—if we give him the blood tubes, he could run it through his chromatography, look at all these metabolites. What he could then spit back was that this was a cerebral palsy kid, this was a Down syndrome kid, or this was an autistic kid, just from those metabolites.

The nice thing about it—yes, pretty impressive, right? But it was unpublished.

Participant: How do we see it? How do we see the data?

Dr. Leshner: People have to get the data into the system.

Participant: I understand that.

Dr. Leshner: I questioned unknowingly before about publication. Particularly in this field I think we have to be extremely careful that we not lead families astray, lead the scientific enterprise astray. So if you could tell Sue Swedo or somebody, people who are actively involved about this, maybe they can get access to the data.

I am a journal publisher, so I am obsessed with peer review and publication and making sure that whatever it is that we communicate is going to be as scientifically rigorous and credible as possible, lest we lead people astray.

Participant: Oh, I understand that. I hate to stand up here and say there is this fabulous tool that has not been worked. But just to throw it out, that it is a tool that has been preliminarily looked at with maybe 100 or so kids. I think it also allows us—the tool also has the fingerprint for the individual child. So what Martha Herbert was talking about, as having a way to track our treatments, I think we should as physicians be allowed the freedom to do whatever we do, because each doctor may be fixing a certain subset of children, but if we don't know, that is the frustrating world of the clinician.

Dr. Leshner: I agree with that. I would just urge you to somehow get the individual attached to these networks that are developing, because we don't want to lose the opportunity if there is something particularly in this. So perhaps you could refer the physician to the networks that are developing.

Participant: That is my frustration as a clinician. How do we know whether what we are doing is valuable and have the time to collect the data that the scientific community—when I went to UC–Davis (University of California–Davis) too many years ago, I know the rigors that science wants, and we just don't have it available, but we are doing something. So while we are doing something, we might as well be collecting a yardstick so we can measure what anybody is doing, and we don't have that kind of biomarker. But I think this is a potential tool.

Participant: One model that might apply here, and maybe we can talk about it more tomorrow, is the cystic fibrosis (CF) model, in which they started with a few very focused research centers, encouraged them to begin the training. That very rapidly got out to regional centers, and the regional centers began to work with the private physicians, and they markedly improved survival rates for individuals with CF.

So I would be very thrilled to try to help spearhead that effort. Obviously it is going to be a major undertaking, but I think working together we can probably get that done.

Participant: The bloods are already being collected so you don't need a whole lot of blood.

Participant: Right. We would need to make sure that the diagnoses were accurate and blinding was done. So I think the testing of that particular hypothesis, absolutely, that can happen very quickly. The larger question is how physicians can be providing feedback and families can be providing information. It is something I think we need to organize both sides in at the same time.

Participant: I am a grandparent of two autistic grandsons. I thank you for inviting the public. I hope to be one of the taxpayers and voters who gets you the funding to go on with your research.

I must say, when I saw the advertisement for this on the Internet, I was distressed to see that mercury was not going to be discussed in the context of vaccines. I thought there was going to be a white elephant roaming around in the middle there, and everybody would avert their gaze. But I see that there is frank discussion on that, and I am very encouraged. I think the pursuit of science, obviously you have to go where the truth leads you.

Also, as a taxpayer funding this research, I think it is very important for you to understand that the people who are going to be out there getting political want some basic questions answered. They want you to look at the mercury hypothesis and tell them why it is not mercury, why it is not repeated environmental insults, and the number of shots they get. It is mercury and it is an immune assault to get these inoculations, and there are so many of them. So they need an explanation for why that is not the cause.

I think we can't move beyond and do good research until we answer that question, put that one to bed. So let's not ignore it. Let's address it head on, and tell parents why they needn't feel guilty. There are so many emotional issues involved here.

Another thing is, people don't really trust their government all that much anymore, CDC, FDA, and beyond. They need to trust their government more in their research by knowing that certain areas are not off limits.

If we do not allow an explanation for mercury and vaccines, it will be like doing lung cancer research and saying, but let's not say cigarettes. Nobody would believe it, and they would know their money was wasted. So that needs to be on the table.

I have another tack altogether. Psychology doesn't seem to be represented here. I know that is perhaps not a good fit with environmental issues, but it is something that should participate in any funding for autism.

In our situation we have gotten a lot more bang for the buck with the biomedical. We have pretty much given up a lot of the behavioral therapies. They have been good, but they haven't been as good as the other. So we need guidance.

Then one more thing. I have a natural experiment to suggest. Rho-negative mothers are subjected to a standard of care that calls for a RhoGAM injection at 5 months of gestation, so we give them a little extra environmental assault and a little extra mercury there.

I understand that Rho-negative mothers have a higher percentage of autistic children than others, so what is the explanation for that? Is it Rho-negative mothers? I don't know. These are some questions we would like answered.

Thank you very much.

Participant: I would like to comment on the RhoGAM issue. There is a small study which is not very well done, and it shows this kind of finding. Since then there has been a more systematic study which I hope is in press. I can't give more details, but it is from somewhere very respected in the United States, which has done a large sample in a study of that kind, and it showed there was no association between RhoGAM and being the mother of an autistic child. It is in press.

Participant: Just going back to something both Mark Blaxill and Mark Noble said before, we are looking at first of all a very complex set of circumstances here. I don't think there is going to be one thing that is found. I think it is going to be multifactorial, and I think this group is saying that.

But I think as we address that from a clinical point of view, we have to look at not just one treatment modality, but also what the child is experiencing overall. When we look at what Jill James was showing us between the gut and the immune system and the brain, looking at all of those factors before we say, how does chelation affect a child?

I think as we set up the studies that Sue Swedo is doing, for instance, if we set it up in such a way that we are just looking at the effects of DMSA on a child without looking at whether that child has had and still has gut or immunologic abnormalities, we would have very different outcomes than if we look at a child that is otherwise healthy and then looking at how chelation does it.

So it is a very messy set of data and we have to look at all those parameters going in and coming out, or else we are going to have data that show us nothing.

Participant: My name is Heather Elias. There is a subgroup of children with autism, and a critical factor in treating them is regulating their hormone levels, particularly their testosterone levels. We know that boys are more likely to have autism than girls, but the girls tend to be more severely affected by the autism. That also makes me think that testosterone plays a big effect on how these children's treatment should go.

We know that Risparitol, Lepran, Spironolactone, all pharmaceutical drugs, are effective treatments for certain behaviors associated with hormone levels. I am curious if any of the scientists here are doing any kind of research on the hormones and how it affects the behaviors of people with autism and regulating those, how it helps them, if there is any kind of study going on about that.

Participant: The study that we will be describing tomorrow, one of the domains of research that we are investigating, is related to hormone abnormalities.

Dr. Leshner: Good, that will make you come back tomorrow.

Participant: It will be including also immune dysfunction, which will address a lot of the other issues that have been raised today.

Participant: There is someone in the United Kingdom whose name is Simon Cohen, who has been doing studies in relation to autistic symptoms or traits in the children who are born from these pregnancies. He has shown some relationship. It is very preliminary. It is not looking at autism as a disorder, as an outcome.

Participant: I just also wanted to mention, I have a daughter who has autism, and we have followed the DAN protocol, and I am so grateful for these doctors doing all of this research. But we now are looking at the data. Just keep in mind that the parents have basically spent thousands of dollars to get these tests done, and they have really sacrificed a lot to get that information. So just remember the children when you are looking at the data.

Thank you.

Dr. Leshner: That is what this is about. You can be the last word.

Participant: My name is Scott Bono. I live in Durham, North Carolina. I thank the panel for convening and taking up this topic. It is deeply personal. Last month I filed to retain guardianship of my 18-year-old son. I have two other children in college, and I never expected that when my son was born I would have to do this at age 18. It is very personal. I know what happened to my son was inexplicable, but you all are looking into it, and I appreciate that.

The most relevant question that has been asked today was asked by Dr. Choi. That is, for 15 years nobody has been looking at the urine and the blood of the children that we are talking about right now.

Most parents, when they go to a pediatrician and they are told that their child is autistic, they are dismissed. That child's illness is dismissed on the basis of behavior. I am so grateful that each of you is looking into some of the systems that have gone wrong in my son and so many other children.

Treatment should always be what is in your mind as you proceed here, because I want my son back. Everybody wants their child back. I really want to thank all of you for coming. And Jackson thanks you.

Dr. Leshner: Thank you, sir, and thank you for reminding us. We are going to stop for the day. This has been from my perspective a wonderful day. I want to thank the speakers, I want to thank the audience. This was for me a very—I guess the right word is dramatic, but it was a wonderful example of how the patient and family community and the scientific community can work together. You have to come back tomorrow. I am really tough, so if you don't come back tomorrow I am going to chase you.

But I didn't want to leave without making the comment that I am not sure that I have seen as good an interaction between the scientific community and the patient and family community, and I really appreciate it greatly. I am very grateful to the family members for coming and sharing your experience and your insight with us.

Day 2
April 19, 2007

WELCOME AND INTRODUCTIONS

Dr. Alan Leshner

Dr. Leshner: Yesterday was a terrific day, and therefore that puts great pressure on all of us to make sure today is an equally good day.

One of the things it was characteristic of is a wide variety of people sharing their views, but also listening carefully to each other. That is what characterizes a good workshop, that it is actually a workshop and not a bunch of monologues where people tell you just how it is.

One of the things, since I am somewhat outside this field, that I have been struck by is how many research opportunities and how many research challenges exist. That is both good news and bad news. The bad news is that I believe that we are way behind where we ought to be, that there is a tremendous need to have a focused research agenda, but also from my perspective the resources necessary to implement that research agenda as well.

I have been very impressed by the number of NIH Institute directors, the Deputy Director of the National Science Foundation, and a variety of other people who have access to resources who have spent so much time with us, and I applaud that. I applaud their leadership in helping to move this field forward.

Now, just to remind everybody of the ground rules, first, speakers, there is a little clock up here.

The format for the discussions is that after the 15-minute talk, we will have hopefully 3 to 5 minutes for urgent questions of the speaker. We will restrict those to the participants at the table. At the end of each session, we will have an open discussion. Again, we will give priority to people at the table, then if we have time open it to the broader audience. Then at the end of the day, we have reserved a substantial amount of time for discussion by everybody in the room.

Let me now introduce Henry Falk, who is the chair of this first session on environmental epidemiology, using population-based studies to isolate the environmental causes of autism. Dr. Falk is the director of the Coordinating Center for Environmental Health and Injury Prevention at CDC.

Session V
Environmental Epidemiology—Utilizing Population-Based Studies to Isolate the Environmental Causes of Autism

Dr. Falk: Thank you very much, Alan.

There was a lot of discussion yesterday about issues we will be talking about this morning, so I think there is a lot of interest in these sessions. We will start with Irva Hertz-Picciotto, who has a Ph.D. in epidemiology from the University of California–Berkeley. She was on the University of North Carolina–Chapel Hill faculty for 12 years, and is now at UC–Davis Department of Public Health Sciences. Her research interests are in environmental exposures, pregnancy outcomes, and epidemiological methods. She is on the editorial boards of the *American Journal of Epidemiology*, *Environmental Health Perspectives*, and *Epidemiology*, and was on the scientific advisory board for the U.S. EPA. Thank you very much, Irva.

ENVIRONMENTAL EPIDEMIOLOGY STUDIES: NEW TECHNIQUES AND TECHNOLOGIES TO USE EPIDEMIOLOGY TO FIND ENVIRONMENTAL TRIGGERS[15]

Dr. Irva Hertz-Picciotto

Dr. Hertz-Picciotto: Thank you. I am going to provide an overview of environmental epidemiology and epidemiology generally. I'll talk about different study designs and what we have learned from them, and then make some recommendations.

The first couple of study designs I am going to go through somewhat quickly, because I think most of the meat is really at the end, in terms of the future for the field.

Starting with focused clinical studies, these are generally self-selected populations or a group of patients in a clinic. These studies are descriptive. They usually have small numbers of subjects in them. Sometimes the hypotheses are generated a posteriori. These are the

[15]Throughout Dr. Hertz-Picciotto's presentation, she may refer to slides that can be found online at http://www.iom.edu/?id=42469.

studies that have taught us about sibling recurrence, about twin concordance, male–female ratio, the comorbidities, and the genetic syndromes that seem to also go sometimes with autism, as well as seizure disorders and gastrointestinal symptoms. They have taught us about the heterogeneity of onset, including the regression phenomenon that seems to happen in a lot of cases, and the data on anthropometrics, such as head size, have come from these studies as well.

The second kind of study, described in the next slide, is based on administrative databases. These are large databases that are collected for administrative purposes. The diagnosis of autism is frequently done by whoever the clinician is who happens to see that child. We have learned about perinatal factors and about time trends from these studies. There are two ways in which these studies assess exposure. When exposures are not assessed in the individual and the outcomes are summarized at the group level, for instance, by area, it is called an ecologic study; in the other design, both exposure and outcomes are assessed at the individual level.

This is an example of a time trend study conducted in Denmark using an administrative database of diagnoses. It was a study looking at the removal of thimerosal from vaccines and the rates of autism before and after. What you can see from this time line, which was not necessarily obvious from the original paper, was that before thimerosal was removed, there was a period of time when only inpatients were in the database, and during part of the "after-removal" period, which covered all the way out to 2000 in this study, there was an interval when both outpatients and inpatients were included. This study, therefore, is not a rigorous design, because as you can see, you can't really compare the before and after periods because of artifacts in how the database was constructed, and specifically, in how that changed over time.

The next slide shows another administrative database study, quite a good one done from the Swedish Birth Registry, which was linked to their inpatient register. It gave us some information about some of the aspects of the perinatal period that seemed to be associated with higher risk for autism.

From these administrative database studies, we have learned about patterns: We have learned about the age effect of the parents, obstetric complications as risk factors, and aspects of the time trends in autism.

Moving along to the genetic studies, these again, like the clinical studies, are volunteer samples. The largest such study right now is the

Autism Genetic Resource Exchange (AGRE) database. It has over 1,000 families right now. They are all multiplex families with at least two members who are affected, and they are focusing on genetics. We have learned from these studies how highly concordant monozygotic twins are, but not entirely. We have learned that this is not a condition that follows simple Mendelian inheritance. We have also seen the slide show yesterday by Isaac Pessah and others depicting the large number of chromosomes that may be involved, indicating multiple genes. But at this point those studies have focused on genes in isolation, although Clara Lajonchere has been working with me to figure out how to collect more environmental data on the AGRE families.

Just a few words about studying environmental factors in autism. I think there are some misconceptions, maybe not in this room, but certainly out in the field of science and the community at large. We know from the genetic studies that about 60 to 90 percent of cases have some genetic component.

What can we conclude about environment? Let's look at these two pies. This is the sufficient causes model from epidemiology. Each pie represents a set of sufficient causes that will cause autism in at least one individual out there. It might be that A is a genetic factor, that B is another gene, that C is an environmental factor in the prenatal period, and D might be something happening at birth or postnatally, just hypothetically.

Looking at the lower pie, this is another set of sufficient causes, where A and C can be substituted with some other set of events. Each set, that is, each causal pie is sufficient. So what that means is, if you take away B from either one of those pies you don't get autism.

This is gene–environment interaction. Because of gene–environment interaction, environment plays a role in 10 to 40 percent at a minimum. But notice that in this individual, say someone corresponding to the upper pie, it takes both. Let's suppose that this particular set of sufficient causes produces 30 percent of the autism cases. That means 30 percent require genes and 30 percent require environment. Suppose the other pie corresponds to the remaining 70 percent of autism cases: Under this scenario, 100 percent of the cases require genes and 100 percent require the environment. In other words, the contribution from genes plus the contribution from environment do not have to sum to 100 percent, and they will not sum to 100 if there is any gene–environment interaction.

There are also several environmental factors that have been associated with autism with very high relative risk. The first is congenital rubella. In the mid-1960s, the United States experienced an epidemic of rubella. Mothers who had rubella during a pregnancy and passed it transplacentally gave birth to children at a much higher risk for autism, about 10-fold higher, and that figure is based on counting only the cases that did not seem to resolve over time.

Thalidomide: also, a very high relative risk. Just looking at a few other factors such as maternal age or male sex, that are not necessarily causal, but might be proxies for some causal factors, there are also some large relative risks.

The fourth type of study design I will talk about is the new generation of case-control studies. These are population-based studies where the diagnosis is confirmed in all of the individuals. They cover a broad range of factors. Generally speaking the exposure has been assessed retrospectively, but that is not 100 percent true. If you go back to get medical records from these individuals, you are collecting essen-tially prospective data, data that were originally written down in a prospective manner. These studies also have been collecting specimens with linkage to laboratory scientists.

I'll give an example. There aren't very many of these studies, but you have heard yesterday a little bit about the Norwegian study, not a case-control study, but they are doing nested case-control studies within it.

This is a case-control study we are conducting in California. CHARGE stands for Childhood Autism Risk from Genetics and the Environment. We currently have about 800 participating families. They include children with autism, children with developmental delay but not autism, and children from the general population. Each child with a potential diagnosis for autism is assessed with the ADOS and the ADI and other assessments are done on all children, including cognitive and adaptive development, a physical exam, medical history, and a structured interview that takes about an hour and 40 minutes, covering 12 domains. Six of them are shown here, including information about the index pregnancy, household product use, metal exposures, and so forth.

Then there are some self-administered forms about comorbidities and also about treatments and services that the child receives. We collect urine, blood, hair, and we ask the mother to bring in the baby hair lock if she saved it, and many of them have. They don't have to give all the hair;

we can do with a few strands. We also collect specimens from siblings and parents, and then we go back to get medical records. We have them sign medical record release forms. It's a very labor-intensive process, and we try to obtain as many of the types of medical records shown on the slide as possible.

In addition, in California there is a banking of newborn blood spots on every newborn. Currently we have about 480 dried blood spots and we are applying to get more.

Overall, this is the scheme that we are working with. On the left side is a panel of broad classes of exposures that we drafted as priority exposures to take a look at, and on the right side are the methods for collecting data about those exposures. Just a few examples: From blood, we can measure pesticides, and we can ask about what pesticides were used in the home. We can also link the residential information the mother provides with some databases that are also available in California, which have a record of every commercial application of a pesticide anywhere in the state. The database has geographic locations of applications, which we then can link to the residence of the mother at the time of birth, or at any other time because we collect those residential data.

Measurement of metals can be conducted in blood, hair, the baby lock, the newborn blood spot. We ask about fish consumption and other household product use for metals. Another example, data on infections can be abstracted from the medical records and is collected by interview from the medical history. You get the picture.

The CHARGE study is in progress. With regard to other case-control studies of this type, there is also an autism phenome project, which Sue Swedo talked about yesterday; it involves NIMH and the M.I.N.D. Institute. The CADDRE study we will be hearing about later in the session from Diana Schendel.

Just briefly, some of the things that we are starting to see: In CHARGE, we are finding very different immunologic profiles in the children with autism as compared to the control groups. A wide range of immune markers appear different in the children with autism.

We have also examined gene expression, and have observed a set of genes that seem to be differentially expressed in the children with autism, especially in cells from the immune system called the natural killer cells.

There are also some hints now that there are perinatal factors, potentially avoidable ones, that might be linked to autism. We have looked at the metals and last year reported at the IMFAR meeting those results. This year I will be reporting on PBDEs; that work was funded by Cure Autism Now, which is now in the process of merging with Autism Speaks.

The CHARGE study, by the way, is funded by the National Institute for Environmental Health Sciences as part of one of the children's centers. It began in 2001, and we are now in our second 5-year period.

The last study design I wanted to talk about is the prospective cohort studies. These are studies where we start with a pregnant woman, and follow her and the child forward. We have already heard a little bit about the National Children's Study. A couple of other studies are now looking at high-risk cohorts. In particular, the pregnant women are ones who already have a child with autism and are carrying another child. Because of the high sibling recurrence rate, these are high-risk pregnancies. One of these studies, which is also part of our Children's Center, is called MARBLES; we are also collaborating with the EARLI network, which is scheduled to be funded beginning in 2008, for which Craig Newschaffer will be the PI (principal investigator). What we are focusing on is trying to find out how early we can see biological signs of autism. Several baby sib studies to date have been focused on the early behavioral, but not biological, indicators, and have started postnatally.

Our aim is to determine what are the critical time windows for environmental exposures, and what are the biomarkers that we might use to identify pregnancies and children at high risk. However, any marker is only useful if we can additionally identify what are the target issues or receptors or enzymes on which we might intervene in order to interrupt the development of autism.

In conclusion then, we have had a couple of decades in which we have learned a lot about autism from psychologists and psychiatrists. We have learned how to diagnose reliably, and even some early behavioral markers. We have learned from neuroscientists about aspects of the brain and brain growth, and now is the time for the environmental epidemiology and toxicology to work together.

This is an area that has not received a lot of attention. What you see in this room is what is out there, pretty much. We have in this room about 80 percent of the environmental epidemiologists and toxicologists, who are looking at autism from this perspective. And only half the scientists in this room are currently doing autism research. So this is an area that has been understudied. There are very few studies of environmental factors in the causation of autism. What I have shown you is basically about what there is.

These large epidemiologic studies can be linked to all the mechanistic questions that were raised yesterday. We are doing genomics and there is a potential to do metabolomics and proteomics to figure out biologically, physiologically, molecularly what is different in these children. With the prospective studies we can take the hypotheses that we are getting from the case-control studies, such as the finding from the CHARGE study of differences in expression of certain genes, and ask: Do those same genes differ, and how early on do they differ? Does it start at 18 months, 12 months, 6 months, or what about in the cord blood? In other words, let's go back to the very early stages of life and brain development. That is the state of the art and that is where we should be going.

Thank you.

Dr. Falk: Thank you very much, Irva. There is time maybe for one or two very quick questions.

Dr. Fombonne: Your design includes data that looks at a range of exposures, but you do not include in your model any timing of exposures. A lot of data suggest that early exposure is important. Therefore, my question relates to how you can modify your study design to look for common exposures at early time points.

Dr. Hertz-Picciotto: In the case-control studies, obviously that is the deficiency in the design of the case-control study, is that you are measuring things now, and you really want to know what happened before the diagnosis, perhaps in the prenatal or perinatal period.

As I said, the medical records do provide us with early information. We have been looking at even preconception, looking at things like in vitro fertilization, medications that the mother took during pregnancy, what kind of induction or augmentation of labor happened. So there is that component.

We looked at the metals so far in the concurrent blood samples, and now we are going back and taking the baby lots. We started measuring the metals in the baby lots, which represent exposures in usually the first year of life, and then the newborn blood spot, which will tell us something about right before the time of delivery.

So yes, it is a problem. In the questionnaire we also asked whether we can get valid information. It is subject to how well people can remember what happened, and do cases remember better than parents who have a child who is developing quite typically, if you ask them what pesticides did you use around the home when you were pregnant or in the first year of the child's life.

Obviously that is a problem, and that is partly why we are moving now to also doing prospective studies, where all the information will be collected prior to the diagnosis.

Dr. Fombonne: May I ask another question? There are also techniques to look at clustering around different environmental exposures. There is currently a study in California looking at increased risk of autism to exposure to different pesticides at early periods of development and gestation. They have found really interesting findings that you can look at one particular exposure and map that to autism rates.

Dr. Hertz-Picciotto: Yes, that is quite a lovely analysis. That is using administrative databases and then linking them to exposure databases. Where those databases exist, I think that is a really excellent approach.

One of my graduate students has been doing—it is not looking to exposures, but she is looking at spatial clustering in California, and breaking it down. But yes, those exposure databases are quite useful, and the pesticide use report one is a very interesting database.

Dr. Falk: Thank you very much. The second speaker is Craig Newschaffer. He will speak on environmental exposures in autism international studies. Craig is professor and chairman of the Department of Epidemiology and Biostatistics at Drexel University School of Public Health. He had founded and directed the surgical office in development, disabilities, and epidemiology at Johns Hopkins previously.

ENVIRONMENTAL EXPOSURES IN AUTISM: INTERNATIONAL STUDIES[16]

Dr. Craig Newschaffer

Dr. Newschaffer: Thank you. I have been tasked to talk a little bit about the promise and potential of international studies and international epidemiologic studies that shed some light on environmental exposures and autism.

This framework of looking at frequency by person, place, and time predates the formulation of epidemiology as a discipline. It guides us still

[16]Throughout Dr. Newschaffer's presentation, he may refer to slides that can be found online at http://www.iom.edu/?id=42470.

in doing basic descriptive work on disease frequency.

What we find, however, is just describing frequency of disease is important for assessing pharmacology burden, but very quickly we transition from just a pure description to wanting to make some inferences about risk factors or causal mechanisms. In person, place, and time, the variation of disease across those three domains is very important to analytic epidemiology. The dimension of place plays a prominent role in this, and there are a few ways that place factors into this type of thinking. For the rest of my talk—place—I am going to be focusing on national variations. As Eric brought up, we can also talk about studies of place looking at smaller geographic units, but again, for the rest of my talk I am going to be focusing on cross-national variation.

The way we bring these data on place into our thinking is, there is a classical application that I will be talking about in a little bit greater detail that in theory can help us shed light on whether or not environmental or genetic factors may be prominent in etiology for the particular disease under study.

Then there are also location-specific opportunities. We heard a lot about these yesterday, studies taking place in specific locations because there are higher or lower exposures in those particular locales. I will talk about those a little bit, too.

Then lastly, the notion of place in analytic epidemiology in terms of multilevel analyses. These are the sorts of analyses where we look at variables at the individual level and also at the contextual level or level of place. A variable like socioeconomic status (SES) in modern studies of epidemiology is often looked at from a multilevel perspective. My personal socioeconomic status might influence my risk for a particular outcome, but also the SES in the area that I live might have a separate and independent effect on outcome.

I think that at this point, where we are with autism epidemiology is probably not to the point where multilevel analyses are going to be prominent, but I think the first two approaches are worth talking about in greater detail.

What about this classic application? The first thing that we want to do when we want to think about variation in place or variation cross-nationally in terms of helping us think about environmental versus genetic causes, we need to rule out bias. We need to make sure that variation in measures of disease frequency in one country and another country reflect underlying risk and aren't related to other factors.

I think all of us in the room are somewhat familiar with the discussions about time trends in autism prevalence. While we may have differing opinions on where the level of evidence on that takes us, I think we all acknowledge that in thinking about time trends in autism, there are some difficulties that emerge in trying to figure out whether or not prevalence from one time period is truly different in terms of risk, reflecting risk from another time period, or other related factors such as changing diagnosis and recognition. The same kinds of issues are going to arise in cross-national variation studies.

The next thing we would do after we rule out bias is, we need to think about the extent and the magnitude of variation across countries. In general, if we see a lot of variation across countries and the variation is of great magnitude, we feel that that suggests a probable role for an environmental cost.

It is theoretically possible that genetic variation could bring about large differences or multiple differences across countries, but typically when you see these patterns, they are associated with environmental risk factors. However, just looking at the extent and the size of the variation isn't enough. What we do in epidemiology, we look to some other fairly simple designs to help us follow up on international variation patterns.

One common design is to do migrant studies. In migrant studies we look at individuals who move from one nation to another and see whether or not their risk is different from individuals who stay in the home country.

There are challenges here as well, because if there is a risk difference we need to determine whether that is because of external environmental factors, exogenous chemicals in the environment that they have been exposed to, or it is because they are adopting lifestyle changes of the new nation. That can be a challenge to sort those things out as well.

For autism, migrant studies are going to present some challenges. If you are talking about cancer, a disease with a long induction period, individuals can move from one country to another and their risk profile can be altered. For autism, we are talking about a relatively small induction period, and there are going to be challenges to migrant studies.

We also look at ethnic variation studies. There, we look at different ethnic populations in one locale and look to see if there are differences in disease frequency across those ethnic populations. If there are, this can suggest that maybe there are genetic mechanisms in play. But again, complications. There are also sociocultural factors that move with ethnicity, so it can be difficult to tease those effects out as well.

We also have to remember that we have got to be cautious in interpreting ecologic associations based on international data. If we use resident-to-proxy exposure, we can run into some problems with inference, and I'll give you an example of that in a minute.

I just want to show you some data for prostate cancer. A lot of good data on incidence rates of prostate cancer and other cancers cross-nationally. Here we see almost a 100-fold difference from China to Northern America on incidence, but even for cancer, where the data are very good, the interpretation is somewhat complicated. We see that on the mortality side, the difference isn't that great.

We know that there are large differences in recognition of prostate cancer across countries. Prostate-specific antigen screening is much more common in developed countries than in developing countries. We also know that there are ethnic variations here in the United States, with much larger incidence rates in African Americans than in Caucasian populations. But even though we see this large pattern, we suspect that there are environmental factors involved in prostate cancer. Interpreting these data is challenging.

Here is an example of the problems we get when we try to use nation to proxy specific exposure. These are plots of breast cancer incidence on the y axis, and on the x axis the countries are ordered by an estimate on how much dietary fat intake there is in the country. The lower countries on the left, and higher countries in terms of fat intake on the right. This is averaged data.

When we do this, we get a very nice correlation which suggests there might be an association between dietary fat intake and breast cancer. But we know from individual-level studies that there really has not been consistent evidence that this association exists. Yet the group-level studies, the national data, have sort of misled us. So we need to be cautious when we do these.

Location-specific opportunities. Higher-background-exposure outcome prevalences are going to give us more power to detect associations, comparable associations. We heard examples of that yesterday. This is one reason why we might go after studies in specific locations.

Here are some examples from the literature on polycyclic aromatic hydrocarbon exposure in fetal growth. The cohort at the left is from Krakow, Poland. If you just look at the bottom two rows, you see that the proportion of exposure through food intake is comparable, but the personal exposure measures on the moms is much higher in Krakow. That is because of airborne exposure to PAHs there. That is a high-exposure cohort.

This is another example from lead. We can see a high-exposure cohort on the left from Australia. If you look at the lead concentration data on the bottom, you will see that there is much higher exposure there. You will also see that there are very few low-exposed individuals in the Australian cohort on the left.

This is something that we need to be careful about if we do these kinds of studies. If the exposure is too high, we lose variation, and we might not be able to see all the effects that we want to when we reduce that variation in that way.

The other reason why we do targeted studies is because of different genotype prevalences. If we suspect that genotype prevalence is higher in one nation, we might want to focus on gene-finding investigations there, because with similar logic it will give us enhanced power to identify those genes. For genotypes that are very rare in one location, if they are more common in another, it may help us find the full range of genotypes that are involved.

In gene–environment interaction, the situation is a little bit more complicated, and kind of interesting. If there is gene–environment interaction, the exposure main effect may be more detectable in populations where the susceptibility is higher. So if we go into a population that has higher prevalence of susceptibility genotypes, we may be able to find the environmental exposure easier. The same logic applies for gene finding. If we move into countries where exposure levels are high, we may be able to find the susceptibility genotype more easily, and we might be able to estimate the actual interaction between the gene and the environment more easily in those situations.

Now I want to talk a little bit about what we know about autism. We have been interested in international variation in autism for even greater than 25 years, as this slide implies. There are case study data going back 30 and 40 years.

These are data on autistic disorder prevalence by time and also by location. If we take a quick look at this, we see that in the earlier years, there basically were limited differences across regions. There is a cluster of Japanese studies that were a little bit higher in the mid-80s, and that has been attributed to the fact that they were using DSM-III criteria and had very intense case findings.

If we look into the more recent studies, we see a lot of variability, although it is interesting to note that in the very recent studies, variability seems to be narrowing, around 30 per 10,000. Remember, this is for autistic disorder prevalence, not the full spectrum. But if we look at patterns by region, we don't really see anything striking. But what is striking is that these are all developed countries, for the most part. We have limited data so far in developing countries.

In part through the efforts of CDC and Autism Speaks, investigators who are doing autism research internationally have been brought together in a series of symposia. What I am showing now for the remainder of my talk reflects my take on what has been presented in those symposia.

What we note first is that there are now studies beginning in developing countries, as indicated by the hatched dots. The studies are of a variety of different designs. We have some freestanding prevalence surveys, some freestanding etiologic risk factor studies, but what is probably most common is this combined design, where there is some population that is going to give us prevalence estimate, but then there is also some recruitment of a non-case population so that some etiologic risk factor studies can flow from this.

In this combined design, there are different ways that case finding is done in different countries. Some case-finding approaches rely on registries and service system records, and these tend to be in the higher income developing countries, and others rely on population screening, and these tend to be in the developing countries. This can cause some problems with bias. The type of case identification approaches can make for noncomparability across these studies.

I am going to have to move a litter faster because I am cognizant of my time.

From work that we have done on case finding, where we have been doing some screening in a developing country, China, we have come up with some issues. We found some language problems in terms of adapting our English screeners to the Chinese population, and we have come up with some cultural problems related to gestures being discouraged and persistence being highly valued in that culture. Some objects used in common screeners don't exist over in China. So there are some real challenges in adapting screening and case-finding techniques to developing countries.

To conclude, moving forward, I think we are going to see more disease frequency studies in developing countries. That is going to be motivated in part by the developing countries' concerns to characterize the burden of autism. Autism still has profound public health significance even in countries where there are other child health problems at higher prevalence than the United States.

When we do this though, we need to make sure that we can maximize comparability of methods, and we can't characterize the differences that are used across developing and developed countries.

I think the initial analytic studies that we see in developing countries are going to be much like the initial studies in the United States that Irva just described. We are going to be looking at a broad array of environmental risk factors. I think that we are going to need stronger leads from basic science, including toxicology, clinical science, and other epidemiology, before there is justification for doing special focused analytic studies in particular locations where there is high exposure prevalence.

I think gene-finding studies are going to go on in developing countries. They are most likely going to be primarily motivated by this notion that there might be some important functional variance that we just don't see in the West that we might see in other populations. But I think it is important for us to remember that if gene–environment interaction is important, there is added motivation for doing these gene-finding studies in developing countries that have a very different exposure profile, because we might be more likely to see the genes that work in concert with that environmental exposure.

In my last 24 seconds, I want to acknowledge the folks who have arranged the international epidemiology symposia, Autism Speaks and the CDC and the participating investigators there, and my colleagues on our China pilot, because I wouldn't have gotten into the international studies business if it weren't for all them.

Thank you.

Dr. Falk: Thank you. Remember that we will have time for general discussion after this session, but if there is a quick question, we have a minute or two.

Dr. Martinez: Since I am not in the field, I read your recent article just published, "Summary of the Epidemiology of Autism." One of the things that you stressed there is something that I referred to in my talk, which is the difference between ethnic groups in the United States, for which the same problems of interpretation that you have mentioned are there, and it is that precisely.

Since we have representation of an underdeveloped country in the United States today, which is from Mexico, a huge population, a very good school-based study, to try to confirm that this is due to bias as most people interpret it, would be very interesting.

Dr. Newschaffer: I totally agree. He is referring to the results from the Hispanic population from the CDC surveillance study. There are a number of locations in the network in the United States where there were significant differences between the Hispanic and the White population, but there are also a number where we didn't see significant differences. I think it is very important for us to focus and understand this, to see whether it is real or not.

Dr. Falk: Thank you very much, appreciate it, Craig. The next speaker is Diana Schendel, who will be talking about the CADDRE study in environmental epidemiology. Diana is a lead health scientist in epidemiology, team lead in developmental disabilities branch of the National Center on Birth Defects and Developmental Disabilities at CDC, and serves as science liaison for CDC Center for Autism and Developmental Disabilities Research in Epidemiology, the CADDRE project, and is principal investigator for the department in metropolitan Atlanta.

Thank you, Diana.

ENVIRONMENTAL EPIDEMIOLOGY STUDIES: CADDRE[17]

Dr. Diana Schendel

Dr. Schendel: Thank you. I will be describing for you today the programs at CDC that can meet the challenge of environmental epidemiology and an epidemiologic focus. These programs were initiated by a congressional mandate for CDC to establish autism surveillance and research programs following the Children's Health Act of 2000.

The programs that were initiated include the autism and developmental disabilities monitoring or ADDM Network, the Centers for Autism and Developmental Disabilities Research and Epidemiology or CADDRE network, and the "learn the signs, act early" campaign. But I

[17]Throughout Dr. Schendel's presentation, she may refer to slides that can be found online at http://www.iom.edu/?id=42471.

will be focusing my comments on the ADDM and CADDRE networks.

The ADDM Network goals are to establish a comparable population-based estimate of the prevalence of autism in various regions around the country, to describe the characteristics of children who have autism, and to examine these trends over time.

This map illustrates the locations of our ADDM sites. The light blue sites are those represented by our current ADDM grantees. Georgia colored in yellow represents CDC as one of the participating ADDM sites.

I can't describe for you the ADDM methodology this morning, but I would like to highlight some of its strengths. All the ADDM sites apply a common case definition, and the majority of sites also can apply a uniform case identification approach. As the map illustrated, we have multiple sites throughout the country, and consequently the ADDM Network population base represents a very large proportion of the population of the United States. At our peak, the ADDM sites represented a population base of around 10 percent of 8-year-olds in the United States, and 8 years old is the target age range for our monitoring program. And of course, the ADDM Network is intended to be ongoing in order to monitor prevalence over time.

This may seem like a far cry from what is needed in an environmental epidemiology framework, but in fact, the ADDM Network provides a very fundamental role by providing us with a much better understanding of the patterns of occurrence of autism as well as the patterns of occurrence associated with some very broad environmental factors, such as variation in geography or community and by sociodemographic factors.

The ADDM data can also provide some very important understanding of the impact of certain methodologic factors on the variation in the observed prevalence that we see. This is just one illustration of these features.

This is a representation of the ADDM Network data for the study year 2002. It is the prevalence of 8-year-olds in 2002, and the overall prevalence was 6.6 per thousand. But as you can see across the different study sites, the variation of prevalence varied considerably, from a low of around 3 per thousand to a high of around 10 per thousand.

These data also illustrate, though, an important component attributable to methodology, most likely. That is, it did seem to vary according to the type of source, where the information on the cases was being obtained. As you can see in the group of sites on the right-hand side of the slide, these sites had unrestricted access to both education and medical records to identify their cases with autism. For those sites, the prevalence variation tended to cluster fairly closely around between six and seven per thousand.

The sites on the left of your slide had either restricted or no access to education records, and were relying only on medical records or medical sources for information, and there the variation was much broader, ranging from around three to seven per thousand. So clearly, it seems to me that the number and types of sources where you are getting information on children with autism has a considerable impact on the observed prevalence.

Therefore, the ADDM data can provide very important baseline referent data that can be used for comparison with other studies, whether it be to examine trends over time or across space, which might be used, for example, in specific locales who might be looking at the impact of particular intervention or prevention programs on the prevalence of autism, or in communities who have a concern over an apparent cluster. The data in the ADDM Network can also inform public health policy by providing numbers to quantify the public health burden of autism in a given community, as well as identifying vulnerable, at-risk, or perhaps underserved populations, all of which are important reference points for developing a framework for environmental epidemiology.

Finally, the ADDM data can provide clues, as Craig Newschaffer has already described, regarding potential broad environmental factors. For instance, if we see social or economic class gradients, demographic or geography gradients, or ethnic gradients, these may be markers for an associated environment factor, that is, assuming that we can be confident that the variation is not attributable to some methodologic bias.

But as we have seen considerably in this workshop, environment is a term that encompasses a very complex mixture to disentangle. Typically we are identifying single or a very few number of components within this complex mixture, whether they be from the physical or biological environments, or whether they might be impacted by an individual's lifestyle or behavior. But in reality it is much more complex, where these individual components don't act alone, but are combined into causal pathways leading to disease.

In autism we may be actually looking at multiple outcomes, multiple disease outcomes referring to the different autism subgroups, all of which may have different associations with a particular causal pathway.

Finally, the genetic underpinnings of autism may be contributing to this variability, as we have seen, if the genetic component changes an individual's susceptibility or resistance to a particular environmental factor.

Stepping into this enormous analytic need that we have in environmental epidemiology is the CADDRE program, contributing a small part to helping us disentangle this complex mixture. The CADDRE program in its first funding cycle was charged with developing a collaborative epidemiologic study that was conceived at the time to be offsetting a deficit in federal programs with an explicit population-based epidemiologic focus. The product of that effort was to develop a study to explore early development or SEED, which I will be describing for you now.

This map illustrates the locations of our CADDRE sites around the country. It includes five study sites as well as a data coordinating center site, and Georgia serves as the sixth study site for SEED.

The main research areas in SEED include investigation of the autism phenotype not only to better understand the unique features of the autism phenotype, but with the objective to identify specific subgroups for etiologic investigation.

Our two main etiologic domains are the roles of infection and immune function and reproductive and hormonal function in autism. We are looking at GI features, not only as a component of the autism phenotype, but as its potential role in the etiology of autism. Clearly we are looking at genetic features, especially those related to our primary etiologic domains, and we are looking at sociodemographic features, not only as their potential role in etiologic pathways in autism, but also as a part of the process of identification and diagnosis of children in the community.

Some of our other areas of interest include the impact of lifestyle behaviors in pregnancy. We have been looking at select mercury exposure, such as maternal vaccine exposure and RhoGAM exposure in pregnancy, and child vaccine history. We are looking at the occupational histories of both parents during pregnancy, sleep features of children with autism as well as patterns of hospitalization and injury.

To do this we have adopted a case cohort study design, and we will be using a population-based approach to identify three groups of children: a group of children with autism, a neurodevelopmentally impaired comparison group, and a group called the subcohort, which represents our general population.

This slide simply illustrates very generally our study design. Our base population consists of children who are both born in and reside in each one of the current study areas. Within this base population, children who are being served with developmental problems in the community serve as our source of children for the case and NIC groups, and the subcohort are drawn randomly from our base population.

This slide illustrates our identification process for our case and NIC children. We are casting a broad net to identify these children. In other words, our eligible diagnosis includes not only autism, but a broad array of other related developmental problems. This broad net will include both children with an autism diagnosis as well as those with one or more of these other developmental conditions, and all of these children will be administered an autism screen.

Children who fall above or who achieve a score above our cutoff for the autism screen will be designated a possible case, and those who fall below the cutoff will be randomly selected to serve as our NIC group participants. I might also add that the target age range for children in this study is children ages 24 to 60 months.

Then our subcohort, as I said, was randomly selected from our base population, most of whom are typically developing, but nevertheless all of these candidates will be also given the autism screen. So for the purposes of enrollment, they are considered members of the subcohort, but if the child scores above the cutoff, that child will be sent through the data collection comparable to a case in order to have a full clinical evaluation.

When we went to estimate how many children we might be able to identify through this process, we applied a very conservative prevalence estimate of about three per thousand to our estimated base population of about 485,000 children in our study areas, which gave us an estimated total number of children with autism of 1,550.

Following potential losses due to inability to contact or ineligibility for the study or for refusal to participate, we hope to enroll around 900 children and families with autism. In following the possibility that not all of these families will complete the full data collection, we hope at the end to have at least 650 children and families with autism in our study. Since we intend to enroll the NIC and subcohort groups in a one-to-one ratio with cases, that means we will be enrolling around 2,700 families and getting full data collection, we hope, on 1,950 of these families.

Based on this projected sample size, we did a calculation of the minimal detectable risk we might be able to detect with adequate power across a variety of scenarios of different exposure prevalences in different autism subgroup sizes. As you can see over a broad range of exposure prevalences, we have the ability to detect risks of between 1.5 to 2, although clearly at the most rare exposures and smaller subgroups, the power of our study declines accordingly.

Our data collection includes a variety of interviews and self-administered questionnaires, including a 7-day stool diary and 3-day diet history. We will be doing extensive medical abstraction, similar to what was described for the CHARGE study, including preconceptional records for certain providers of the mothers, if they had psychiatric or immune dysfunction or hormonal dysfunction, and we have 3-year postnatal medical record examinations of the children.

The children will be given a full clinical evaluation and a brief physical exam, and we will be collecting biologics from both the child and both parents, including hair from the child.

We hope with the biologics to be able to measure both proteins and genes as well as mercury in the hair samples. Our list of possible candidates for analysis is not fixed at the current time in order to accommodate new discoveries in the meantime until we are ready for analysis. But the data collection protocols, of course, will accommodate a variety of analytes to be tested. We have also included in consent to allow the retention of sample after analysis for further testing in the future.

Clearly these analytes that we choose to measure will primarily address our key etiologic domains in terms of brain function or immune or hormonal function, but we also hope to measure epigenetic or parent-of-origin effects, all of which we have learned may modify or be modified by the environment.

Given these strengths, we believe SEED has a role in environmental epi of autism. Certainly since it is the only multisite study planned to date representing diverse communities and populations, it certainly will increase the generalizability of our study to the U.S. Since we are following a population-based approach to identify our subjects, this will enhance the representativeness of our sample and reduce biases due to subject selection.

Since we chose to have two comparison groups, an affected and unaffected group, we hope this will enhance our ability to detect biases due to differential recall between parents with and without an affected child, and therefore reduce our exposure misclassification. Since all sites will be applying a uniform protocol, we can pool data across sites and therefore give us the largest sample size of a study planned to date, which clearly will enhance our study power and permit phenotypic subtyping for etiologic analysis, which although that does reduce the sample size we would be using, we hope it reduces imprecision due to outcome misclassification.

Finally, with multiple research domains, we can look at multiple causal pathways or look at multiple points within a single pathway, and with some overlap of other studies that have been described, we can replicate analyses from prior studies.

So we think the primary role of the SEED study for an environmental epidemiology perspective is that it simply gives us a much better understanding of the role of a variety of broadly environmental and genetic factors in ASD which can serve as a referent for studies that might be focused on more specific toxicological factors. We have the potential to expand the array of environmental factors that we have data on through data linkages or added data collection. Clearly it can also serve as both a hypothesis-testing and hypothesis-generating study and inform data collection and analysis in other studies to come.

In terms of a recommendation, we have lots of needs, but probably the greatest need is the fact that a single study can't do it all. The data collection that is needed and the analytic approaches are simply too diverse to be accommodated by a single study. So perhaps one area we might focus on is data pooling or coordination across studies. By coordination I don't mean centralized control, but perhaps arriving at some consensus on certain data elements and analysis which can improve the strength of meta-analysis or data pooling for *de novo* analysis.

Thank you.

Dr. Falk: Thank you very much. We do have several minutes, if there are any questions at the moment.

Dr. Beaudet: I'd just like to ask, does the CDC have any initiative to try to understand if the incidence of autism changed from 1970 to 2000? It seems like if it did change and we are at some new plateau, it would be nice to know that it did change over that interval.

Dr. Schendel: We do, although we don't have the data sources here in the United States that would address that question. But we are working with investigators in Scandinavia and Denmark, which do have these historic registers, and we are attempting to use those registers to investigate the impact of a variety of administrative changes in diagnosis or referral of children in those areas, to see what that impact might have had on the change over time.

Dr. Beaudet: But it seems to me like I could interpret your answer to say there is no effort to answer this question for the U.S. population.

Dr. Schendel: We don't have the data. We only started the ADDM Network beginning with the 2000 study year.

Dr. Beaudet: But there are some things that could be looked at retrospectively. If terbutaline administration is a high-risk factor, it seems to me that by retrospective study could be elucidated.

Dr. Schendel: Where would you get the autism prevalence data?

Dr. Beaudet: I guess I would be thinking about some kind of case-control study of autism with controls in terms of exposure to terbutaline in utero. I'm not an epidemiologist, so feel free to point out my ignorance.

Dr. Schendel: That certainly would give you a risk estimate, but it is not going to give you an estimate of the number of children who might have had the disease in the general population.

Dr. Falk: This might be an interesting issue to further explore during the general discussion session.

Mr. Blaxill: Dr. Schendel, there is a lot of confusion about rates and trends and comparability of various statistics. Why would you make the choice to lump all of the autism spectrum disorders together in a single measure, as opposed to distinguishing between the different subclassifications? That strikes me as a very unfortunate choice.

Dr. Schendel: Well, that is true, there are components within the autism spectrum. Even though we do have the *DSM-IV* criteria, which may give us some differences within the spectrum of different subtypes, it is still not clear how that may be well implemented in this kind of study to where we are confident that the subtype that we have assigned a particular child based on behavioral data reported in the record would be reliable.

So to be conservative, we have chosen to lump the diagnoses across the spectrum. Although we do collect a variety of behavioral data, as I said, it may be difficult and would have to be validated, with the ability to subtype.

Dr. Leshner: So let me ask you a question about that. If you did that in your study, but if the database were widely available, could other investigators see if they could parse it out?

Dr. Schendel: The data are there yes, to the extent that these data are recorded in the medical records of these children. We aren't examining these children. We collect what information is available, and that information is not perfect.

So while we feel confident that within the broad spectrum we can identify a child who may or may not have autism, it is much more challenging to take that information and reliably subtype. So is the direction we have taken. But the data we have are there.

Dr. Leshner: I'm just asking whether it would be possible for other people to access that database.

Dr. Schendel: You could. The data are available. It would have to be validated through a clinical validation study to confirm whether or not you can make a correct judgment call on that data.

Dr. Newschaffer: Just to clarify for people who might be listening and confused, there are two studies. They have now transitioned to talking about the ADDM surveillance study, which is records based, not the SEED study, which is going to involve direct examination of kids.

Dr. Schendel: Right, you're right. Maybe that is the confusion. The SEED study will have the clinical validation.

Dr. Leshner: I was thinking of CADDRE.

Dr. Newschaffer: CADDRE will be doing extensive subtyping.

Dr. Schendel: Thanks, Greg, I didn't catch that confusion. The SEED study will have clinical validation of all the children, and we will be able to do subtyping based on the SEED study.

Dr. Newschaffer: We are taking everybody in the spectrum in, but we will be able to subtype, including conventional and also nonconventional ways, by incorporating other information on medical presentation and comorbidities and symptomatology. So it will be extensively subtyped.

Dr. Falk: One more question from Sallie.

Ms. Bernard: But Dr. Newschaffer, the SEED study is not going to give us prevalence, is that right?

Dr. Newschaffer: Correct.

Dr. Schendel: Right.

Ms. Bernard: So knowing the subtypes—we lost it on ADDM, and if we are getting it through SEED it is really not delivering what we need to get.

Dr. Newschaffer: That's right, yes.

Ms. Bernard: My other question for you, since we are not getting the information that we need from our U.S.-based studies, are you getting subtypes in your international studies, and are you getting different age groups, including into the older population, since we are just focusing on very young children in the United States?

Dr. Newschaffer: First of all, they are not my international studies. If we look at the international studies across the board, there is a lot of heterogeneity in terms of what they are doing. Some are looking at different age groups; some, especially those that are based on more developed countries might be able to do subtyping. In developing countries where they are doing screening and have to come up with a way to validate diagnosis, none of them have gotten far enough. I suspect they will face some challenges in terms of doing subtyping.

So I think it is going to be a mix for awhile as these international studies develop.

Dr. Susser: Can I just ask that you redirect us to this question in the discussion period, because I think it is important.

Dr. Falk: Yes, I was just going to suggest that. Thank you, Ezra. We will come back to this discussion for sure.

The last speaker in this session is Allen Wilcox. He will be speaking about prenatal and perinatal exposures. Allen is the senior investigator in the epidemiology branch at NIEHS, where he has worked since 1979. He was chief of that branch for about 10 years and serves as editor-in-chief of the *Journal of Epidemiology*. He is also past president of most of the major epidemiological societies in the United States.

Thank you, Allen.

PRENATAL AND PERINATAL EXPOSURES[18]

Dr. Allen Wilcox

Dr. Wilcox: Thank you for the invitation to be here. I am not an autism researcher. I thought maybe it was a mistake when I first got this invitation. But I am glad to be here. I have learned a lot.

I am going to start by diverging from my text. The reason is, I would like to address a question that came up yesterday that I think is an

[18]Throughout Dr. Wilcox's presentation, he may refer to slides that can be found online at http://www.iom.edu/?id=42472.

important one, about the categorization of autism, and looking for categories of phenotype that might help us understand causation better.

There were a lot of discussions about ways to do that, using molecular biology, biomarkers, signs and symptoms, but no one mentioned epidemiology. So I would like to introduce the idea that epidemiology can actually help construct categories of this outcome that would be useful for unraveling its cause.

I am going to present an analogy from another area of epidemiology, birth defects, and specifically oral facial clefts. This type of birth defect comes in what might seem at first to be a bewildering variety of manifestations: it can be only a cleft lip, it can be only a cleft palate, and it can be various combinations of both of those things.

What we have come to understand is that these are actually two different birth defects. One is cleft palate by itself, and the other is cleft lip with or without cleft palate. The first recognition of that was by family studies by epidemiologists who showed that a family who had one child with a cleft lip had a 25-fold risk of having a second child with that same defect, but their risk of having a child with a cleft palate was by comparison much, much less—much closer to the general population. The converse is also true. The families with cleft palate have an enormous risk of having another baby with a cleft palate, but the risk of having a baby with cleft lip is much closer to the general background level.

Since that time, we have come to understand the embryonic origins of these two defects and why they are distinct. Understanding that they are distinct has led epidemiologists to study these two defects as separate categories. This in turn has been very important for understanding their causation. I can give you an example.

We did a study looking at folic acid and how it might prevent cleft lip or palate. We found that for cleft lip or palate, there was a 40 percent reduction with folic acid supplement, whereas for cleft palate there was no association. If we had combined all of these birth defects, all of these facial defects into one category, we would have had a much harder time identifying this association.

There is an analogy in autism. You have the opportunity in multiplex families, in studies looking at recurrence of autism, to see whether there are different categories of autism that tend to cluster together in families. If there are, these might be used to create categories of autism for etiologic studies. I'll just put that on the table for your consideration.

Now back to our regularly scheduled program. I want to discuss the possibility of prenatal and perinatal origins for autism. I am not discussing the important questions of treatment. My comments have entirely to do with the discovery of causes that are presentable.

The fact is, we know very little about causes of autism in general, and of prenatal causes in particular. But I would like to go through a number of arguments why looking in the prenatal period might be a productive line of research for autism. I am also going to talk a little bit about the difficulty of doing these kinds of studies, and finally, I will say something about the opportunities for doing better than we have done in the past.

Plausibility. Some of these things are very familiar to you all, and have already been touched on by various speakers. We know the fetal nervous system is particularly susceptible to toxins at a low level of exposure that would not affect adults. The exposure to organic mercury at Minamata is a prime example of an exposure that had little or no effect on adults, and yet had devastating effects on the fetuses who were exposed.

Thalidomide was a very important discovery for epidemiologists and clinicians because the exposure was not merely benign to the mother, it was actually prescribed to the mother. This severe birth defect (phocomelia) was the result. And as Irva Hertz-Picciotto has mentioned, phocomelia is associated with autism.

So we know the fetal nervous system is particularly susceptible to toxins. We also know that the fetal brain has a very long period of susceptibility.

This figure is from an embryology textbook by Moore and Persaud. Along the top you see the whole sequence of prenatal events from conception through embryonic and fetal life, ending with the term infant.

Each one of these colored bars represents a particular organ system and the period of time during which it is vulnerable to being disrupted. The red bar indicates the period of time when there can be structural malformations produced by exposures, and the yellow bar is the period of time in which functional problems can be introduced.

You may have trouble seeing it, but the top bar is the central nervous system. As you can see, it is the only organ system that is vulnerable from the earliest embryonic life through delivery—and beyond, as we now know.

Does this mean the fetal brain is fragile? I think we have an important responsibility in our public health role to find the right balance in our communications to the public. On one hand, we know there are thousands of chemicals that are potential neurotoxins in our environment. We know the fetal brain is very susceptible. On the other hand, we know that the brain is, in the language used yesterday, plastic. It is robust. It can manage to accommodate insults without being damaged. The fact is that most babies develop normally, despite all the things that their mothers may worry about. As experts in this field, we have to find the right balance, not to frighten the general public, but to create an awareness of serious issues that we need to address scientifically.

A third point about prenatal exposures and autism is the interesting association of autism with difficulties of pregnancy. Much of this evidence is quite old. The studies go back to a time when research methods may not have been as good as we would expect today. But there is a sense that pregnancies that eventually result in an autistic child have some signs during the pregnancy that things are not exactly right. These are nonspecific, and I certainly would not want to suggest that the signs and symptoms that you see on this list "cause" autism. However, they may share some common cause with autism. These studies provide one more sign that the roots of autism, or at least of some cases of autism, may go back into the prenatal period.

Whenever you hear epidemiologists talk, they will always tell you how difficult their work is—I am going to be no exception. There are some things about autism that make it particularly difficult for the kind of studies we would like to carry out. Autism is a rare outcome. Autism is hard to define. The exposures that may be important could also be rare, and they are often hard to measure. An example from yesterday was the discussion of the "hit-and-run" exposure, that is, a fluky exposure at a crucial time. How do we ever go back and reconstruct that?

The retrospective reconstruction of exposures, as a number of the previous speakers have commented on, is always difficult. This is one reason prospective studies are particularly useful—they can gather exposure information in more detail, and we can then try to associate those exposures with the emergence of the disease later.

The difficulty with prospective studies is that, for a rare disease, those prospective studies have to be very large. Until recently, this was not considered practical. However, the new wave of large prospective pregnancy studies offers new opportunities.

We have heard some wonderful presentations this morning about promising studies of various creative designs. I am going to talk about the one that I happen to know best, which is the Norwegian cohort. You have heard this mentioned several times already in this meeting.

In the cohort, the mothers are enrolled by week 18 of pregnancy, with a target population of 100,000 mothers and pregnancies by the end of 2007. I think they are close to that target. A great advantage of doing this kind of study in a medically well-organized country is that there are many registries and routine mechanisms of collecting health data which, when combined with universal identification numbers of all citizens, allow information to be assembled in a way that would be difficult and expensive in the United States. The plan in the Norway cohort is to follow these offspring through age 6, but of course longer would be better.

This kind of a cohort study is a very ambitious undertaking, even in a place like Norway. It is expensive to do, and it has no immediate product. So there has to be some sense of building an infrastructure that can be used later for useful things. Let me tell you about my own institute's participation in the study. This is one of those stories that starts out, "there were these two guys in a bar." I was in Oslo having a beer with Per Magnus, who was one of the principals in the Norwegian cohort. This was 6 or 7 years ago, when the study was just on the verge of going into the field. The Norwegians had plunged into the study, with no clear idea of how they would pay for its completion. My conversation with Per Magnus went along these lines: You have great opportunities to measure pregnancies and pregnancy outcomes, but you're not paying much attention to environmental exposures. Environmental research was not the Norwegians' expertise, not something that they had much experience with.

So I suggested that I would try to get my institute to provide some support to this cohort, specifically to pay for the collection of biological specimens during pregnancy that we could use to measure exposures. Specimens collected during the time when the fetus was actually developing could be stored and later assayed for toxicants that might be relevant to particular poor outcomes.

I went back to my institute, and the leadership of the institute thought this was a great idea. We were able to provide money for the Norwegians to collect urine and blood during pregnancy. Blood is being stored as whole blood and plasma. We are prepared to assay at least this list of potential environmental exposures. To measure some of these things, for example heavy metals, we have to collect specimens in a very careful way in order to avoid contamination after the sample is collected. But we have done this. The samples are now in storage. I notice that a number of these exposures were on Irva Hertz-Picciotto's list of hypotheses that she had for autism.

More recently, Ezra Susser was in my office talking about this great project that he was involved with, in which they planned to identify autistic kids in the Norway cohort. He and I immediately recognized that this was a perfect partnership, in which we could provide biological specimens for measuring specific exposures and they would find the cases. So we made plans on the spot to combine our biological samples with their quite complicated and complete study of autism in Norway, all within the framework of this cohort study.

So my pitch here is that these cohort studies provide a structure that can be useful in ways that we don't even anticipate when they are started.

Thank you very much.

Dr. Falk: Thank you, Allen. If there is a quick question for Allen we'll take that now. If not, we will go into the general discussion. So perhaps we will move to the general discussion.

Questions or discussion points?

Ms. Bono: I have several over the course of several speakers. For the ADDM surveillance, Diane Schendel, you were talking about, you were going to be looking at trends. But then Art Beaudet asked a question about birth cohorts. I know that you have looked at the 8-year-olds back to the 1992 and 1994 birth cohorts, but I'm not quite sure much further back. Then you were saying that those data weren't available.

I am concerned about that, because you are looking at education in, but in the education sources there have been a couple of studies of the 1989 and 1992 birth cohorts, where autism jumped. So if that is the case, we need to be going back further and looking at those years and seeing what changed in the environment at that point.

For the SEED study, I was wondering if there were going to be any nonvaccinated in the study. I see that you have got 1,950 enrolled children by the end, but if you are looking at exposures, are there going to be any kids with no exposures to vaccines? Because that is certainly an issue that we need to deal with. So that is number two.

Number three, the mercury and hair samples, I just want to remind everyone that there is a baby hair study, where the one thing that they noticed in the study is that the children that had autism had very little metals in their hair. That was the determining feature of those children. So once again, that theory of the nonexcreters. You may not see mercury in the hair of the kids that go on to develop autism, and that may be something you want to think about.

I think those were the main points I wanted to make.

Dr. Schendel: At CDC we have for the 1996 study, we have children born from 1986 to 1993, children ages 3 to 10 in 1996. But that is just limited to—it is not the higher cohorts of the ADDM Network. The ADDM Network, our first study year was 1992 births for children aged 8 in 2000, and we are moving forward.

I don't know whether the ADDM Network has discussed going back further and getting earlier cohorts, but currently they are moving on to the 2004 and 2006 study cohorts.

Ms. Bono: It would seem to me, if the jump was earlier, then we need to go back earlier. We need to make that a priority in this room, that if the jump in autism was in 1989 and 1992 or perhaps even earlier, that late 1980s jump, that is when we need to go back and figure out what happened in the environment, because then the trend just continued.

Dr. Falk: One of several comments we have had in terms of looking over a longer period of time.

Ms. Bono: Right, it is of huge importance in time trends. So to look now at a prevalence, it is very important; we still want to look at these kids, but we also need to go back and look in time.

Dr. Falk: Thank you. There are two more questions. Do you want to say anything about the other questions?

Dr. Schendel: We are not specifically targeting children with or without vaccines. It is children against everybody within the study areas, some which may and some which may not have vaccine history.

Ms. Bono: That begs the question of whether we need to look at that, because that is certainly an issue that we have heard over and over, and mercury is implicated throughout this entire workshop.

Dr. Schendel: Maybe in the CHARGE study. Dr. Hertz-Picciotto, you have a specific psychological focus. Perhaps you could describe what CHARGE is doing.

Dr. Hertz-Picciotto: We have been trying to reconstruct all sources of mercury exposures in the children in CHARGE, using diet, using use of household products like nasal sprays, vaccines, RhoGAM, and something else I cannot remember right now.

So we are trying to ascertain all the sources of mercury in the child's environment and then trying to correlate that with whether they have autism or not, and also related to the blood mercury levels.

Most of the children—since the mercury was taken out of most of the vaccines, although it is still in some vaccines, there is a difficulty in ascertaining whether people have been exposed to thimerosal in vaccines, in particular influenza vaccine, which may or may not have thimerosal, depending on where they are getting it. Although the medical records are supposed to record lot and batch numbers, we have not always found that in the medical records, for one thing.

The other problem is that people who now get these flu vaccines at their local mall, there may be no record that we can determine. So in other words, we may be able to ascertain the vaccines they have gotten, but we may not always be able to determine—some of them we can exclude thimerosal because thimerosal is no longer used in those, some of them we can't always determine the dose the child would have had.

So this is what we have been running up against in trying to do that analysis at this point in time.

Ms. Bono: I just wanted to also comment regarding the limitations of that. Can you tease out non-exposure to vaccines, a zero-exposure category? Also, there is an overwhelming viral load as well in these children. So if there is a way to tease out all exposures to virus through vaccines and naturally, I think it is important to include too.

Then also, we still don't know the amount of trace exposure to thimerosal with vaccines, because it is used in the processing. We know there are low-level exposures, and we don't know that that is safe either. So these are all things that I would love to see outlined in the study

Ms. Redwood: I wanted to follow up on that same thing. I guess I am a little concerned with the study, in that if you are asking the question of vaccine exposure, there is not going to be enough variability. As you know, they are mandated by law; children have to have them to get into daycare centers and schools.

As we saw in the VSD data when the first analysis went through, there was a zero-category exposure. If you are looking at all the children being exposed, you are not going to have enough variability. You need a zero-category exposure, which we had in the VSD data, and we saw statistically significant associations. But then the entrance criteria were changed to where all children had to be vaccinated to be in the study, and it took out the zero-level exposures and the statistical significance decreased.

So I would like to see appear a vaccinated–unvaccinated study. I think as an advocate from the community, that is what we hear from parents over and over and over again.

Dr. Schendel: I appreciate that, and I certainly understand there is a definite concern. When we planned the SEED study, it was at the same time that CHARGE was being planned, and there were other activities investigating vaccine safety at CDC. So we chose not to focus on that particular etiology, rather go at the main etiologies I described.

So I think what you are saying is important, and I think it should be responded to. As I said in my last slide, we need multiple studies. SEED will address one area, and we need complementary studies to look at the others.

Ms. Redwood: But that is mercury and aluminum. If you look at the full vaccine load, you are looking now specifically at mercury. That will include aluminum. Over 90 percent of the flu vaccines this year administered to pregnant women, infants, and children contain thimerosal. Maybe that will help you in your analysis.

Dr. Newschaffer: Something else to consider in terms of the age of entry for a retrospective study is, the older the kids are, self-report information, even medical record information, becomes increasingly compromised. So for other sources of mercury exposure like diet or questionnaires and the types of things that Dr. Hertz-Picciotto used, there is a trade-off when you get older kids in with recall. That also affects other exposures that are of interest, too, where that is our prime source.

Dr. Hertz-Picciotto: In regard to the issue of children who weren't vaccinated, that is true; very few children in this country are not vaccinated at all.

Recently, the NIH held a workshop specifically on the vaccine database that CDC has acquired, which is a better way to look at the vaccine question than either the CHARGE or the CADDRE SEED study, because it is large enough to have children without vaccines. We did make some recommendations about improved studies coming out of that large database, the vaccine safety database.

So as I understand it, there are some studies that are now being done, and there were recommendations for other studies to be done. I think that is probably the best place for that particular hypothesis to be addressed.

I think your point is well taken, that when there is a limited amount of variability in exposure, then there is a limitation in how much you can learn from that. Those data in one of Dr. Newschaffer's slides were making that very point. So thank you.

Dr. Schwartz: I had a potential suggestion here. As a nonepidemiologist, I am learning a lot about these epidemiologic studies that are ongoing, and some that have been completed in this area. But it is a little bit unclear to me exactly what the strengths and the weaknesses are, what the complementary and overlapping areas of goals are for each one of these studies, and what the gaps might be that need to be filled by studies that have not yet been funded or solicited.

I'm just wondering whether there should be an effort to try to look at all of the studies. Maybe that is what is going on in the matrix, the autism matrix. I don't really know, but I guess I am throwing this out there as a question. Do we need an effort to try to look at these studies that are ongoing and have been completed as a way of trying to get a sense of where we are and what needs to be done?

Dr. Hertz-Picciotto: First of all, none of them is completed. In CHARGE we have our first 5 years of data, which include as I said about 800 subjects. You did fund us for another 5 years, and we do expect to get about another 800 subjects. So we will have about 1,600 at the end. Then the CADDRE study will probably be slightly larger than that.

I think Dr. Schendel's point about trying to look at ways to combine data across these studies may be a very good idea. We haven't really talked about that directly, although in the planning stage we did share some of our instruments so that there would be some similarities in the data that are collected. As you can see, the designs are very, very similar.

So I think that is a good point about potentially getting more statistical power, particularly if we are looking at the gene–environment interactions question, which does require large sample sizes.

Dr. Schwartz: But I am really asking a different question here. As someone who gets to weigh in or at least provide an opinion about what should or shouldn't be funded, it is always difficult for me to know what is being done, as opposed to what is the new research that is being proposed to be done and how that complements, overlaps, or meets an area that is not being addressed. And if it is not being addressed, it all of a sudden takes a higher priority in terms of funding.

Is there some attempt to do that among these ongoing studies that haven't yet come to fruition? Art Beaudet pointed out a very important point, which was, what is the incidence prevalence over the past 20 years, and how has that changed?

Dr. Newschaffer: There is another point that you need to consider in this. I know that you are very cognizant of this, being in the position that you are in. It is not only the questions that aren't being addressed; it is the ability of the data and the study designs to answer those questions validly and accurately. That has to be part of the matrix as well. Some of these questions, they are real fundamental challenges to doing that.

Dr. Hertz-Picciotto: Let me give some specific examples. I think we are going to have a very good ability, and we have already started looking at those data, and there are some hypotheses that are arising, some associations that we are seeing, to look at some of the perinatal factors when we can get the medical records, and we are working very hard to beef that part of the project up right now, because it is not easy to always get these records. But I think that is an area that is going to be well covered.

I think between us and the CDC study, that is an area—not that we have all the answers now, but I think in the next 5 years we are going to get a lot of information on the prenatal and periconception and peridelivery part of the period. Those are good data.

Now, when it comes to environmental exposures, as you know, the problem is how do you reconstruct exposures that happened in the past. And questionnaires are one of the tools we have. Some things can be remembered well and some things maybe not so well, and maybe more variable and more subject to the state of mind the person who is trying to remember is in.

What can we measure? In California we have the blood spot that can be used to measure metal levels. We can also measure cytokines. We cannot measure any transcriptome, we can't measure many of the other—we can't measure cell-specific functions in immunology, for example.

So the retrospective studies have those certain limitations in terms of prenatal and early postnatal exposures that fundamentally limit case control studies. Those will be addressed in the Norwegian cohort, the National Children's Study, and the high-risk pregnancy cohorts of mothers in the MARBLES (Markers of Autism Risk in Babies—Learning Early Signs) study in the early network. The problem is there are so many things that are so open, that we don't exactly know where to go at this point.

Dr. Schendel: I guess I want to reiterate the point that I made at the beginning of my presentation. When we conceived the CADDRE program in 2000–2001, as I said, it was still an evident deficit in the number of federally funded projects with a specific epidemiologic focus. With the exception of CHARGE, there were no others. So we started from ground zero essentially, and we have tried to plan a study that was, we thought, capable of addressing some specific questions, but yet broadly, broadly, broadly, broadly conceived.

I do think there is a continuing deficit of these kinds of studies. The Norwegian cohort, although that is not a U.S. study, will fill that. The National Children's Study will fill it, but not any one study can do it. So I think to advance the kinds of questions that you are proposing here, given the challenges in data collection, we do need a more concentrated effort and expansion of these particular types of studies, because one study cannot do it all.

Dr. Falk: Dr. Schwartz, let me try and summarize just for a minute. David and I have had similar kinds of conversations lately over various issues from Hurricane Katrina to global environmental health. The issue for us often is what can CDC do, what can NIH do, where are the gaps?

I think since these studies were designed perhaps 6, 7, 8, 10 years ago, maybe it is worthwhile, as David Schwartz is suggesting to think about what is covered by the current group of studies and what is perhaps missing, and some suggestions to that. That might be worth maybe coming out of this afternoon's discussion with some suggestions about that.

So maybe we can all think about the current studies and in the afternoon discussion we will come to that.

Dr. Leshner: Very briefly, if you can, what I was going to suggest is, listening to this conversation and again not knowing a whole lot about the subject, this is clearly an unresolved gap and one that ought to get included in what we talked about.

Ms. Bernard: I want to bring up gaps. So you are saying wait until this afternoon to talk about gaps in the research?

Dr. Leshner: Well, we have had a long discussion about one gap, but I think there are other gaps that we ought to be talking about more this afternoon. I'm afraid of getting started on another topic and then being unable to give anybody a break.

Session VI
Technology and Infrastructure Needs for Future Research

Dr. Falk: This session will be focused on technology and infrastructure needs for future research. The very first speaker is going to be Larry Needham from the CDC Environmental Health Lab. He will be talking about body burden measures.

Dr. Needham is the chief of the Organic Analytical Toxicology Branch of the National Center for Environmental Health at CDC. He served at CDC for over 30 years in the area of assessing human exposure to environmental chemicals through biomonitoring, authored or coauthored over 400 publications dealing with multiple chemicals, and has been very much involved and is past president of the International Society for Exposure Analysis.

CDC ENVIRONMENTAL HEALTH LAB—BODY BURDEN MEASURES[19]

Dr. Larry Needham

Dr. Needham: Today I will emphasize exposure to stressors, particularly environmental chemicals that may have the potential to lead to autism spectrum disorders.

In 1989, the National Research Council first set forth the concept of the exposure–effect continuum, which traces an environmental chemical or some other stressor from its source into the environment, then to human exposure, and finally to a possible effect. Exposure means contact of the stressor with the individual. The portion of that exposure dose that gets into the body is termed the "internal dose." Downstream from that, a particular amount may go to a target organ, and a portion of that may then become the biologically effective dose. After various pharmacodynamic processes have taken place in the body, adverse health outcomes may or may not occur.

---

[19]Throughout Dr. Needham's presentation, he may refer to slides that can be found online at http://www.iom.edu/?id=42473.

Thus far in this meeting we have talked about various environmental stressors. These can be roughly categorized as chemicals, infectious agents, or social factors. By chemicals, we mean environmental chemicals; so-called "occupational" chemicals (these are often the same as environmental chemicals, but generally are found at higher concentrations in people); pharmaceuticals; personal-care products; diet; and household products.

The individual stressors that we have discussed include genetics and personal characteristics such as age, nutritional status, health status, and sex, and many others.

In addition to the environmental stressors and individual stressors, we have talked about autism spectrum disorders occurring in a population mostly from the confluence of environmental stressors and individual stressors that affect the fetus through the mother or affect the infant directly. Within such a population, after various pharmacodynamic processes have occurred, there may be a range of effects, which may (1) involve no autism spectrum disorders at all; (2) may lead to the development of autism spectrum effect biomarkers, but not to any autism spectrum disorder; or (3) in fact, may lead to cases of autism spectrum disorders. I'll come back to this slide in a few minutes.

These next two slides show the effects of multiple stressors in a series of experiments conducted at the University of Pittsburgh. This slide shows an aquarium, a tank. In that tank researchers put 10 liters of water and 10 tadpoles. Adjacent to that tank was another tank in which a predator (a stressor) may or may not have been present. Researchers added to the water in the first tank the pesticide carbaryl (another stressor) at two concentrations in acetone and just acetone as well, all with and without the predator next door.

Very quickly, the results of that experiment: They found an almost complete survival rate of the tadpoles when only water was in the tank as well as when there was only water in the tank and the predator was in the second tank, as shown here by black circles. With acetone control, there was also almost complete survival of the tadpoles regardless of whether or not the second tank held the predator.

When there were high concentrations of the pesticide carbaryl in the first tank, they found that the tadpoles survived for about 5 days, and then almost all of the tadpoles died, whether or not the predator was in the second tank. At lower concentrations of carbaryl, however, there was a difference. This slide shows the effect of multiple stressors. With just the carbaryl at 0.05 milligrams per liter, there was about a 50 percent survival rate after 10 days, but with the carbaryl at 0.05 milligrams per liter and the predator, there was almost a complete demise of the tadpoles. This very simple experiment showed the effect of multiple stressors.

This slide shows the categories of various chemicals that people may be exposed to. It is taken from a 2006 publication on recent EPA data. I don't know the year of the EPA data. What it shows is that in commerce in the United States, there are 82,000 chemicals. About 40 percent of those are polymers; many polymers, following exposure, are not absorbed by the body. So even in those situations in which people may be exposed to polymers, but the polymers are not absorbed, there will be no adverse health effect.

In addition to the 82,000 industrial chemicals, there are about 8,600 food additives, about 3,400 cosmetic ingredients, between 1,800 and 2,000 pharmaceuticals, and about 1,000 active ingredients in pesticides. That's about 100,000 chemicals.

So where do we start in assessing human exposure to chemicals that may be linked to autism spectrum disorders? At CDC we are interested in biomonitoring, but we also recognize that biomonitoring is not a stand-alone tool, so we also use questionnaire data and historical information. We also especially need pharmacokinetic data so we can interpret the results of biomonitoring, and we should not underestimate the importance of environmental monitoring. In an exposure assessment, we try to combine all three of these approaches with calibrated and validated models to assess exposure to environmental chemicals.

About every 2 years, CDC also publishes its *National Report on Human Exposure to Environmental Chemicals.* This is a copy of the *Third Report* released in 2005. Human samples for the *Fourth Report* are being analyzed in our laboratory now, and the actual document is slated for release in 2008.

Some of the chemicals that we have measured are the very chemicals that we talked about in this meeting. These chemicals include metals, PCBs, dioxins, furans, and organochlorine pesticides. Organophosphorus pesticides have been mentioned here as well as polycyclic aromatic hydrocarbons and phthalates. Some of the newer chemicals in the report include the brominated flame retardants, which include the polybrominated diphenyl ethers; perfluorinated chemicals; triclosan, a bactericide, which is added to many personal-care products; sunscreen agents; and parabens, which are used as preservatives in foods and in other products.

Choosing the appropriate matrix, such as blood or urine, for biomonitoring depends on the chemical, but the choice is also population dependent, involving such demographic characteristics as age, race, and health status. Broadly speaking, there are two primary classes of environmental chemicals. The first is those that are persistent, having half-lives in years. These include dioxins; PCBs; some of the polybrominated diphenyl ethers; some of the fluorinated chemicals, such as PFOS and PFOA; organochlorine insecticides; and lead in various biological stores. For these chemicals, we can perform the measurement now and get a good indication of what the levels may have been years ago if the exposure was somewhat continuous.

At the other extreme are nonpersistent chemicals, which have half-lives in minutes or hours. These chemicals include phthalates; contemporary pesticides such as organophosphorus pesticides, carbamates, and pyrethroids; and volatile organic compounds.

Again, where do we start? This slide was prepared for the National Children's Study. I was the federal co-chair on the Exposure Work Group. We were looking at exposures from preconception to 21 years of age. For studying environmental causes of autism, we are concerned not only about exposures that occur during preconception, but also exposures that occur during the three trimesters of pregnancy, as well as those that occur in infants and toddlers. The point here is that the biological matrix that one would use or that is available for assessing exposure during these different periods varies a great deal.

The Work Group assembled a series of slides to examine various classes of chemicals. This slide depicts monitoring persistent organic chemicals at various life stages. On the left-hand side, we listed the different biological matrixes that might be available to us. Then we looked at various life stages, for example, at adult preconception, each trimester during the fetal period, age zero to 1 year, 2 to 3 years, 4 to 11 years, and so forth, and then ranked the potential for monitoring persistent organic chemicals in these various matrixes at these various life stages.

For example, for adult preconception, we would prefer to assess exposure to persistent organic chemicals in whole blood, serum, or plasma. If those matrixes weren't available, we could still get a good indication of what the levels were in the mother by analyzing the cord blood. If we are looking at exposures during the fetal period, in general we can't get fetal matrixes, so to assess exposure to persistent organic chemicals during this critical time, we can use cord blood or we can use maternal blood, either during pregnancy or after the mother has given birth. Clearly, there are various ways to use biomonitoring during these various life stages for assessing exposure to persistent organic chemicals.

We have a similar series of slides that include nonpersistent organic chemicals, metals, VOCs, and other chemicals. These slides are included in the publication that is referenced at the bottom of this slide.

Again, where do we start? We could begin by listing those chemicals among this list of 100,000 chemicals that (1) are absorbed by body, (2) are increasing in the environment, (3) are increasing in concentrations or in prevalence of detection in people, (4) are neuroreactive, (5) cause oxidative stress, or (6) have structures similar to chemicals that have been linked to autism spectrum disorder.

In listing some of the chemicals that have been linked by various groups to autism spectrum disorders, there is no obvious relation among their chemical structures. Thus, using the similarity of structure for selecting potentially active chemicals may not be a fruitful means of getting to the chemicals of concern; perhaps, we should look at some of these other potential starting points to begin our search for chemicals that may be involved in autism spectrum disorder causes.

We also need to gain information on what chemicals to measure in the body by looking at epidemiological studies. Today we have heard about the nested case-control studies, the sibling and twin studies, and the National Children's Study. CDC is also heavily involved with several of the national children's centers that are funded by EPA and NIEHS, again looking at exposure assessment and effects.

In addition to doing this targeted analysis—one advantage of targeted analysis is greater sensitivity—there are also advantages for casting a broader net. One way to do this, which is in a research mode now, is getting improved resolution by using something like 2-D gas chromatography or HPLC, and then coupling that process with mass spectrometry. Doing so allows us to acquire full-scan information for chemical identification on all the chemicals that are coming out of the chromatographic system as opposed to selecting the chemicals to measure prior to the analysis.

What we would do is take a biological sample and do minimal workup. We don't want to remove any chemicals, neither exogenous nor endogenous. Then we would use high-resolution chromatography, such as 2-D gas chromatography, and obtain a full-scan mass spectra for all the chemicals in that sample. This technique allows us not only to look at individual chemicals on a semi-quantitative basis, but also to look at patterns of chemicals. For example, we could potentially look at people with autism and see whether their chemical patterns differ from people who do not have autism.

We also can gain much from case-control epidemiological studies by working backward. By that I mean, instead of working from environmental stressors down to health effects and if we could come up with some biomarkers of effect in autism cases and work backward on this continuum, then it might lead us to various environmental stressors.

I also want to address how concentrations of chemicals in a particular matrix reflect body burden. This slide shows three routes of exposure and absorption: ingestion, inhalation, and dermal contact. There are very good equilibrium data for concentrations of persistent lipophilic chemicals, such as dioxins and PBDEs, between the lipids in the blood and lipids in adipose tissue and in the lipids in breast milk of lactating women.

For lipophilic compounds, persistent chemicals, we can estimate the amount of body burden by doing blood measurements. For lead, the correlation between blood and bone concentrations is a little more difficult and depends on factors such as age. For fluorinated chemicals, like PFOS and PFOA, which are persistent, there is not an equilibrium between the concentrations in blood and fat because those chemicals are not lipophilic. But there are some data on the equilibrium between blood levels and protein in the liver, so in general we can estimate the body burden for persistent chemicals.

For nonpersistent chemicals, making estimates is much more difficult. There are some modeling efforts that are going on now to help ascertain some of the body burden measurements for nonpersistent chemicals, but going from a urine concentration to the total amount of a chemical in the body is difficult.

In conclusion, biomonitoring techniques are available for assessing human exposure to environmental chemicals; however, we need your help in determining the chemicals or classes of chemicals to measure.

I will entertain some questions. Thank you.

Dr. Falk: Thank you. We have a few minutes for specific questions for Larry.

Ms. Bono: Thank you, Larry. I liked your chemicals in the fourth report that you are looking at blood data. How are you going to be measuring metals?

Dr. Needham: That is in another group. What they do there is lead, cadmium, and mercury are measured all the way down to 1 year of age in blood. Then there is a suite of metals left that are measured in urine.

Ms. Bono: I am concerned about that, especially with this group of kids. You might be able to get it on the population basis, but again blood doesn't show metals with these kids, or urine, unless you use a provocative agent.

I know you mentioned just now protein in liver, which I am not really familiar with that much, or even bone. That gets into more invasive types of things, so there is a limitation there.

Dr. Needham: What we talked about are nonexcreters. So what you are saying there is that neither the blood level nor the urine level may reflect what the body burden is.

Ms. Bono: Just based on thousands and thousands of these kids and what we are seeing, unless you use a provocative agent.

Also, you had a great slide on where to start, and you had the list of chemicals that cause oxidative stress and several things. I think it is important to add to that list what chemicals are injected, because we have a problem right now with the mercury, aluminum, and other additives in vaccines. There are all sorts of other additives.

So when we look at the children's exposures, we also need to look at what is injected as well, because that is going right into their bodies.

Dr. Needham: That would be a route of exposure in addition to inhalation, ingestion, and dermal contact. It would be what is actually injected into the body.

Ms. Bono: Correct. You need to have that injection, very important. Thanks.

Dr. Needham: Thank you.

Dr. Herbert: I want to make a comment. I would like to get back to the discussion, but I want to link it to your slide on working backward from effect markers. I think it is important to think about how we determine what those effect markers may be, the ones that we think are pertinent.

This also ties into my thoughts about the distinction between saying that it is possible that these various substances may affect the organism in utero or during the perinatal period, versus that that is the only time that they are going to affect the organism.

There is also the question of body burdens. These chemicals may come into the organism at a critical period and affect neuronal development, but then they may hang around and keep giving their gift and affect the ongoing ambient environment. That sets up a whole different kind of class of effects that would be missed if you are biased entirely toward looking at altered neuronal development. Whereas some of the variability in the presentation of these children day to day is related to the borderline chemical immunological status that can be affected even by some of the transient exposures that occur substantially postnatally.

So I think that the idea of the effect marker needs to be expanded to take account of that kind of pathophysiology.

Dr. Needham: Okay. I think what you also mentioned was the potential for persistent chemicals especially, but also for nonpersistent chemicals because a portion of those are still sequestered.

For example, the persistent chemicals may go into the fat and then leach into the blood, and what you have is continuous exposure.

Dr. Herbert: Right, over life course.

Dr. Falk: I might just add briefly that the biannual reports are from a representative sample of the United States, and it is a snapshot in a sense. But the laboratory also works at any given time on 60 or 70 different studies with various groups, some of those that Dr. Needham has mentioned, to fill in some of the additional kinds of information over and above the baseline national report.

Dr. Needham: We do collaborate with other investigators with federal agencies, like NIEHS, David Schwartz and his folks, and EPA and others as well, academia.

Dr. Falk: Thank you very much, Larry. The next speaker is David Walt, who is going to be speaking about personalized environmental sensors. David is professor of chemistry at Tufts University and Howard Hughes Medical Institute Professor. He served as chairman of the chemistry department from 1989 to 1996. He has been executive editor of *Applied Biochemistry and Biotechnology*, and is the statistical founder and director of Illumina.

Thank you very much, David.

PERSONALIZED ENVIRONMENTAL SENSORS[20]

Dr. David Walt

Dr. Walt: Thank you, it is a pleasure to be here. What I am tasked to do is tell you what is possible, not necessarily what you are doing today.

I categorize measurements in two general categories: first, assessing what is there, and second, discovering what needs to be measured. This latter category will be discussed a little bit later in terms of being able to identify what needs to be measured and what is technically feasible in our ability to make measurements.

We have a number of different ways of assessing any kind of clinical problem. We can measure the environment for what is present. We have the possibility of using personal sensors or dosimeters for measuring drug doses as well as dosimeters for environmental exposure. During this talk, as we transition toward the future of measurement technologies, we will see a trend toward the molecular level, looking at single molecular lesions, developing nanodosimeters that can circulate in the bloodstream for identifying body burden, and we will also see how we can use molecular methods for presymptomatic and early diagnosis.

This slide shows an environmental sensor that we developed in my laboratory that has been deployed on a buoy. This buoy is just off the coast of Massachusetts, right off Martha's Vineyard. I want to point out that there is a sensor on the buoy that is making measurements. You can see occasional spikes in the data. These spikes are not noise; they are actual spikes of something that is being measured in the environment. If you take a sample of seawater, or for that matter just take a sample of blood and make a single point measurement, you are going to miss all this variation. These data underscore the importance of being able to measure things continuously.

The ability to measure continuously is what defines sensors. We are not talking about sensors when we talk about making a single measurement. We are talking about things that can make measurements continuously. From what I have heard here today, the ability to make measurements continuously is essential.

This next slide shows a sensor used for military applications. It is detecting things that are in the air. These are continuous sensors. They

[20]Throughout Dr. Walt's presentation, he may refer to slides that can be found online at http://www.iom.edu/?id=42474.

are making spectroscopic measurements of things that are in the air path. This system is pretty large, is deployed on a truck, and requires lots of power. If we are interested in environmental exposure and we want to know what is being released, some of these systems are going to have to be implemented in both urban and rural environments to get a handle on what is there, what is their temporal variation, and what the concentrations are in a variety of urban and rural settings. Large, complicated, power-hungry instruments may be one way to get to these answers in the near term and should be considered.

The next slide shows analytical test strips. Such strips are a low-tech method. One can do pregnancy tests this way. We have been performing clinical measurements of renal patients who spit in a tube, dip test strips in the tube, and measure various components in saliva that previously have been measured in the blood. It is a relatively noninvasive method for detecting whether these patients need to come into the clinic and have dialysis.

We are all familiar with glucose tests. This type of measurement relates to one of the points that I made earlier, that is, it is a single measurement. If you eat three meals a day and your blood sugar is at a hyperglycemic or a hypoglycemic level and you take your blood measurement three times a day, you are probably going to miss these extremes unless you just happen to sample at precisely the right time.

This next slide shows what people really want to address the glucose monitoring problem. This system is commercially available and is called a GlucoWatch. This kind of technology can be brought to bear on a wide variety of substances. I am not going to talk much about it, but essentially what it does is, it causes the patient to sweat, and there is a little sensor on the underside of the watch that measures glucose continuously. It doesn't do a great job of measuring glucose accurately but it gives a trend, and it tells the patient when they need to do something, such as take insulin or eat something.

There is a wide range of things we heard in the previous talk that can be measured with laboratory-based instruments. These instruments are getting better and better. They are getting much more sensitive. We can't ignore the fact that while these instruments make measurements of single time point samples, they have the ability to measure lots of things.

Sensors and new measurement capabilities are enabled by new materials. I just want to illustrate a couple of these.

This next slide shows colorful solutions of things called quantum dots. They are tiny little balls with nanometer dimensions. They are presumably inert, although we don't want to use these things in vivo right now because their toxicity has not been fully tested. As a research tool, however, they can be of immense value. As can be seen, quantum dots glow in many different colors. They can be incredibly powerful labels.

Here is an example of three different kinds of cancer cells that were injected locally into a rat, and then three different kinds of nanosensors with different specificity to these three different tumor types were injected in the rat's tail vein. The nanosensors went through the circulatory system and localized very precisely at these three distinct locations.

This next slide depicts molecular science. One can now analyze molecules at the single-cell level. These examples show some of the capabilities that are being brought to bear using nanotechnology.

The next slide shows another application of nanotechnology. These triangles are nanogold prisms attached to a surface. One can create immunoassays with these materials that have the ability to measure as few as 600 molecules. This capability is unprecedented in terms of the level of sensitivity achievable with traditional immunoassay techniques.

Let me now talk a little bit about taking the laboratory and bringing it closer to the patient or to the study subject. These devices are called total sensing systems.

This next slide shows a technology that I think will genuinely enable the field of autism research, and that is new sequencing methods. These sequencing methods can sequence literally an entire human genome in a few weeks for about $100,000. Compare that to $2 billion for a human genome 6 years ago. This trend is going to continue as we move toward $1,000 per genome. If you don't need a whole genome sequence you can measure lots of sequence very cheaply for hundreds of dollars a sample, and you can get tremendous amounts of information about variability, literally tissue-to-tissue variability, as well as even cell-to-cell variability.

This next slide shows a new technology that has been developed over the last decade called lab on a chip, also referred to as microfluidics. All of the capabilities of an entirely analytical laboratory are being integrated onto a substrate about the size of a microscope slide. You are probably all familiar with these slides, about 3 centimeters by about 12 centimeters. These chips can perform many different analyses simultaneously with incredibly small sample volumes, and collect a large amount of data from each sample.

Other promising technologies are listed in this next slide, for example, miniature mass spectrometers. We heard about mass specs already. Now, mass specs are getting to the stage where they can be taken out into the field and perhaps even left unattended to be able to perform analyses on a minute-by-minute basis.

Now let me finish the technology side before getting into needs and opportunities by talking a little bit about two technical advances. One advance is the ability to measure single molecules and the second advance is the ability to measure lots of things simultaneously using array technology.

Arrays can perform lots of experiments and make many measurements simultaneously—a capability that is important in any field, particularly the clinical field. What I have heard today is a lot of bias; that is, you know what you want to measure. I don't think you necessarily want to work that way. You want to measure as many things as you can and use these new technologies to discover things that might lead you to find some new markers—you want to pinpoint some new leads in these kinds of diseases. Technologies exist that enable us to look for rare events, to isolate and identify things that don't happen often, but when they do happen they lead to a problem.

Here is just a very small picture of a type of array that we use in my laboratory. The array itself is about 2 millimeters in diameter. All the spots on this array would take up the entire front of this room. This image is a very tiny section of the array. We can measure hundreds of thousands of different sensors simultaneously. We are not the only ones who can do these types of experiments. This example is where technology is today. You can measure thousands of things every time you take a sample and you don't need a large sample. A single drop of blood will enable you to measure thousands of different analytes in that sample. So the take-home message is that you can multiplex. You can and should want to measure lots of things in every sample.

Here is another platform. This platform measures thousands of things in nearly a hundred samples at once. Each of these little sensors is able to mate with a well in a 96-well microtiter plate with a different sample in each well, so you can measure lots of samples and measure thousands of things simultaneously.

This slide shows some work that we have recently done in the laboratory, measuring single enzyme molecules. I'm not going to tell you how we did this experiment as it was recently published. Each of these little red dots is a single enzyme molecule. We measured the kinetics of each of these enzyme molecules. What would conventional biochemistry predict? It would predict that every enzyme is identical. In fact, every enzyme is not identical. This graph shows the kinetic traces of single enzyme molecules. Why are they all different? Some possible reasons are that there are amino acid substitutions in the proteins and because even identical proteins adopt different conformations. These features give rise to an entire diversity of activities in a population of enzymes. We never see this variation when we look at bulk ensemble measurements. We only see them when we look at single molecules. We talk about lesions when we discuss disease. There are a lot of lesions here. Every one of these enzyme molecules is different from every other enzyme molecule. Some of these lesions may lead to disease, while others may simply be in the normal range of biological variation.

The next slide shows a new technology that has recently been published in *Science*. It is able to look at a single lesion in a particular sequence of DNA with very high specificity—a single molecular lesion! Here is an example of a technology that can detect mutations at the single-cell level.

What are the challenges and opportunities? We have to decide what we need to measure. But what I've tried to tell you is that you don't need to decide all that much, because you can measure a lot of things. I would rather err on the side of measuring lots of things by spending a little more money, and then worrying about what to measure after we have measured a lot of things and got some leads.

Sample collection—I am not going to belabor this point as there are epidemiologists in the audience and they know more about this topic than me. But it is important to remember that when you take small samples you have to worry if the thing that you want to measure is actually in that sample.

Sample types—we can and probably should move toward much easier types of samples to collect.

Better sensor sensitivity is still important and is worth investing in. It is also important to expand the ability to look at many things by using these new kinds of techniques. We still need increased data processing capability and data integration. As we collect more data we are going to need to process it and get information out of the data.

Then there are some ancillary issues such as power for driving these devices, including new battery technology. We also need to consider new deployment strategies such as putting sensors in cell phones, and we need to think about system integration.

It is important to specify the analytes. We can consider using surrogate measurements for the analytes that we cannot measure today. For example, there are things that can be measured easily that might indicate that there is something present that cannot be measured readily. Some sensors are available that can be used today for monitoring. Continuous measurement capability is going to be essential, where you can measure things continuously, rather than take samples and send them to laboratories.

Nanotechnology is going to advance measurement science—it is unavoidable. We are getting better and better at being able to measure things, and because of this ability, it will enable lots of fields, including the field of autism research.

I'd be happy to answer any questions.

Dr. Falk: Thank you. Any specific questions? We will take two or three.

Dr. Wilcox: David, I just want to emphasize a point that you made. In our cleft study we recently used an aluminum panel through CDER to look at 300 genes of 5 SNPs each, and it was very successful, but there is good news and bad news in there. The good news is, we now have genetic information on 1,200 SNPs, and the bad news is that we now have genetic information on 1,200 SNPs.

Our capacity to use all this information that is now possible to generate is way behind the technology to produce it. I know you mentioned data integration there, but I think we can't understate how important it is for the informatics to catch up with this, and we are not there yet.

Dr. Walt: Absolutely.

Dr. Pessah: When you referred to thousands of analytes, were you referring to proteins, DNA, or also small molecules?

Dr. Walt: Yes, protein, DNA, and small molecules. I think it is critical that we address all three of those classes. Those are the three important classes.

Dr. Pessah: So I wonder, when you can go to measuring with validation and with high accuracy and sensitivity hundreds of thousands of analytes, could you reduce the sample population? If we believe that autism is on the individual level, can you gain important information by not going out and getting 2,000 individuals into the study, but maybe 5 individuals in the study?

Dr. Walt: I leave that to others to answer.

Dr. Pessah: It's a thought.

Dr. Hertz-Picciotto: It depends what the variability is in the population in those 12,000 or 20,000, whatever order of magnitude we are at. If there is low variability and we are looking for some signal, it is going to be extremely strong, then yes, maybe five cases and five controls. But most of what we are looking at is probably not going to jump out of this quite in that way.

Dr. Walt: Or at least, that is what we think.

Dr. Spence: Given the urgency question that keeps coming up here, how much does it cost, and how quickly is the price coming down?

Dr. Walt: That is a very good question. One of the things that turns out to be a little bit counterintuitive is, as things get smaller they get cheaper, because the materials cost goes to virtually nothing. When you talk about nano materials, when we make sensors, we can make about a million sensor arrays with a gram of material. A gram of material is a powder in a bottle about like this. That is because we are using microsensors. When you get to nanosensors you can make billions of sensors from a gram.

So I am simplifying it a little bit. The supporting instrumentation that goes with reading it out still has to be developed to a level that matches the capabilities of the small sizes of the micro and nano systems.

Dr. Falk: Let's open this now for discussion of both papers.

Dr. Schendel: I just wanted to follow up with that. For these technologies, how do you know that you have a sensitive and specific measure? I know in some technologies, you may get signals, but you are not exactly sure that the signal is what you think it is.

Dr. Walt: Obviously for any analytical test, you have to test the specificity and the sensitivity. You do that with conventional methods of first testing it in buffer and then adding interference and then going to your relevant samples like blood and urine and whatever and seeing if the measurements that you are taking with those matrixes are any different from those that are in pure buffer.

But I think a more important and more relevant answer has to do with just the advantages of making thousands of measurements. The fact is, when you make thousands of measurements, and this speaks to some of the study that you did with CDER, one data point does not give you an answer. It is looking at the pattern of response that helps you identify what the overall process or the overall change that is going on in that sample is.

So the example is, there are lots of correlations typically when you make thousands of measurements. One thing may go up, another thing may go down, and if you make a number of measurements, you might find that if this one doesn't go up at the same time this one is going down, you know that you have made an incorrect measurement.

So there are correlations within those thousands of analytes that you are measuring that get you away from needing to be as specific as you would be if you were measuring just one thing or two things.

Dr. Needham: His question is related to the last question. As far as sensitivity goes, as you know, David, oftentimes the analyte present in urine or blood is picograms per milliliter or nanograms per gram and so forth.

One thing that you showed was improved sensitivity as a need. So what kind of sensitivity is available for measuring these small molecules?

Dr. Walt: The small molecules, you are talking about organic molecules?

Dr. Needham: Right.

Dr. Walt: Organic molecules, still the gold standard I would say is GC mass spec, which is a laboratory-based type of system. But once you have identified which of those organic molecules in laboratory-based settings turn out to be interesting, then you can spend the time and energy to design sensors that can address either that class of molecules or specifically that molecule with conventional kinds of measurements.

In terms of the sensitivities, as I have showed you, Richard Van Duyne, who is at Northwestern, is using these gold nano particles, these little nano prisms that are on surfaces. He just gave a seminar at Tufts the other day, and he presented data that showed that he can get to somewhere on the order of about 10 to 12 molecules of sensitivity for small organic molecules.

So this is not science fiction at this point, but it is not something that is in the general public domain for routine kinds of measurements.

Dr. Beaudet: I just wanted to point out that the kinds of techniques that were displayed there in the arrays are the kind of techniques that are being used for array-comparative genomic hybridization. These things are in the clinic. They are not in research, they are in the clinic on routine use, and they are not significantly more expensive than other genetic tests. And it is diagnosing tons of stuff.

Dr. Schwartz: I just wanted to point out that what David presented is part of what we call at the NIH the exposure biology program, which is part of a bigger initiative called the genes and environment initiative. It is a trans-NIH initiative that all the Institutes are contributing to, that Dr. Zerhouni supports, and also Secretary Leavitt helped develop.

Francis Collins at the Genome Institute is leading the genetics program within the genes and environment initiative, and NIEHS is leading the exposure biology program, but it is really a trans-NIH program to develop sensors, both personalized environmental sensors and biological responses to environmental forms of stress, that allow us to then take those measurements and move them quickly into studies of populations and individuals that are at risk of being exposed. So NIH is investing in this.

The question I have for you, David, is how is industry investing in this, and what are the roadblocks to have industry invest in this area? It seems to me to be something that would be an obvious opportunity for industry.

Dr. Walt: I'll have to give you a generic answer as opposed to specific examples, but the generic answer is that there are a lot of companies that are exploring new measurement technologies. Many of them are doing things that probably could be relevant to this community, but wouldn't be. So those companies need to be enticed to be able to address problems that have a relatively narrow focus, but that they have to pay attention to those problems because they provide in many cases non-equity funding that enables them to leverage their existing funds, and to do essentially the same thing that they are in business to do, except to focus on a problem that is of interest to a particular community.

So I would say that the biggest problem with getting industry involved is that they have to be enticed to be distracted from their mainstream opportunities.

Dr. Goldstein: I'd like to ask the same question of federal funding, NIH funding, this new initiative which is gigantic and has obvious implications to cancer and heart disease and whatever, how will the autism research community be able to be early in line to be able to use these new approaches? How do we position ourselves?

I think it raises lots of issues that we might go into later in terms of recruitment of talent to make use of this technology and committing to autism research.

Dr. Schwartz: This gets at Tom Insel's question from yesterday. You have 200 assays, but you have 49,000 chemicals out there, how are you going to measure—are these 200 assays good enough at this point to move forward with? Should we wait for the 49,000?

Several years from now, hopefully within the next 5 years, maybe within the next 10 years, we will have assays for several thousand as a result of this initiative. Those assays, as soon as they develop, will be made available to investigators. However, like any assay, even the genomic assays and the genetic variation assays, the SNP assays, come with a price. It costs investigators money to actually do the studies. But we are in the market of trying to support investigators to use those tools as much as possible to apply to the research.

As part of the genes and environment initiative, there is a component in 2009 and 2010 to use these assays in genetic studies, so that we are looking at gene and environment as opposed to just environmental exposures or genetic susceptibility factors.

So if things go the way we plan them to go, in 2008–2009 we will be putting out RFAs to provide funds for investigators to use these environmental sensors in genetic studies of complex diseases. The genetic component of the genes and environment initiative put out an RFA for any complex disease, so I anticipate that autism will be one of the diseases that has at least investigators applying for.

But that is not the limit of the types of studies. Even if autism isn't part of the initial studies that get funded in the genes and environment initiative, an autism study could certainly come in to apply for funds to use these environmental sensors as they become available to look more carefully and critically at populations to see if these environmental exposures are associated with the development or progression of autism.

So I think it is a very open playing field right now, and will remain an open playing field in terms of allowing investigative groups to use these tools to be developed.

Is that what you were asking, Gary?

Dr. Goldstein: I think so. The field has been able to attract really top geneticists who use most top technology to be studying autism. I think we have to anticipate not only the money to do these studies, but the talent to plan and interpret these studies, whether our field has anticipated them, who is going to take advantage of them.

The obvious application this whole 2 days is all about gene–environment interactions which seem perfect. But I think the point was made that probably every environmental researcher in the country is in this room right now, and there are very few of them. So the balance is how we can participate in this.

Dr. Schwartz: I think you are raising a really great point, which is, do we need to think about training as part of what we are considering here? Do we need to think more broadly about the pipeline of investigators that could undertake this work?

Dr. Walt: I want to address both of your comments. Getting to a more specific answer, David, to your question, the companies that are exploring some of these really new technologies are focused primarily on cancer, and maybe on heart disease, because they are huge markets. The ability to detect early stages of CVD, for example, would revolutionalize the treatment and care of that disease, and early detection of cancer may or may not affect outcomes, but at least there is a perception that if you can detect it earlier you can do something about it, and you are going to make a difference.

How do we entice companies that have these technologies to now spend their time looking at environmental exposure for autism and other diseases? Again, it is a tough question. The genetics tools that are available are in fact not particularly relevant. It is an example of technology that has now become mature and is available for any disease. You decide you want to study autism, it is there. There are whole genotyping arrays; there are whole gene expression arrays. They cost you a few hundred dollars in experiment, maybe a little bit more, but not that much more, and the price has gone way down compared to what it was 5 years ago.

So the question is really, how do we draw in those new technologies that can jump-start the studies of a disease without having to wait for them to get mature? There is one answer, and that is money. It is to get their interest diverted from something where they have to compete with a lot of other people to something where there is going to be a market, but it will enable the technology to develop. It is just that they are going to develop that technology devoted to solving an autism problem as opposed to cancer, and then they apply that to cancer once they are a year down the road.

Ms. Singer: Are these technologies being designed to produce data that are compatible with all of the databases that we are investing so heavily in, and the NDAR and the IAN projects and all of the parent registries? Or is it going to be data that are in isolation?

Dr. Walt: I think certainly on the side of the genetic databases, and now there are protein databases, there are issues with validation that are being addressed. On the genetic side they have to be compatible. On the protein side, there are similar kinds of movements that have happened over the last 5 years with the genetics to make sure that new technologies that come into play are somehow correlated to those existing databases.

So I think that those are issues that need to be addressed. It is an important issue, but I would say that the technology infrastructure is just not there yet to have that be as mature of a question as it should be.

Dr. Falk: We have a few more minutes before the break, but this maybe goes to Gary Goldstein as well as to David Schwartz, but in terms of training needs for people working with the instrumentation, but going back to Alan's opening comment, maybe also training needs for people using all the data and informatics and so on. What exactly do you see as the main needs in terms of the kind of people that need to be trained?

Dr. Goldstein: To move quickly, my sense is that the successes in genetics have not been taking people interested in autism and training them to become geneticists, but it has been by creating the incentives and the interest of the leadership in genetics to tackle these problems.

At this moment, in the last few years, the very top of this field has now begun to work on autism. Otherwise we are talking about a 10-year feed-in of young people who might work out. So I think it is a question of how we get the talent that exists already to be convinced that it is worthwhile of their time to work in the autism area.

That will also—these people will train people, and there will be a follow-through for the future. But I think the first step for those who want to see something happen in the next 5 years is to take people with these talents and convince them to work in autism.

Dr. Newschaffer: I think the bioinformatics issues are generalizable across diseases. The data don't know whether it is a CVD or an autism dataset. The training in bioinformatics to address the data reduction issues that Alan highlighted can be advanced-disease independent in many ways, I think. I don't know how to do it, but I think it can be.

Dr. Schwartz: I would just ask the mental health community how many psychiatrists or psychologists are trained in environmental sciences to address issues that are relevant to environmental causes of psychiatric and behavioral problems.

I think there is an opportunity in that. I see that as a deficiency, and probably an opportunity to think about developing programs that could address that need in a very clear, scientific, programmatic way.

Dr. Insel: If that was an actual question, David, the number is probably less than zero. But in terms of the more general question of what do we do, going back to Gary Goldstein's issue, how is any of this relevant to autism, what I am hearing is something that is so similar to where we were maybe a decade ago in genetics. At that point, we were spending an awful lot on trying to genotype relatively small samples of many, many different disorders.

Then about 5 years ago we said, enough of this, we are not getting any answers. Let's just collect the DNA and bank it from as many people as we can, put more of our effort into tool development, getting the capacity to be able to do things like whole genome association, and when that capacity finally arrives, we can then move very quickly.

So I think we are probably in the same framework here. The question that Gary Goldstein asked was the question that I had after looking at the NIEHS website describing the exposure biology program. It is very difficult to tell from there where anybody who has a specific interest in a given disorder will get into that. It is all about capacity building.

Until you answered the question, I didn't understand the overall plan. It is just like genetics. You decided, let's build the tools and then when we are ready in 2009 probably, we will be able to then use them.

But I think now, to go back to Gary's issue, what should we be doing currently is collecting the samples, so that when the tools are there, we don't have to start. We will have the biorepositories. I think the discussion to have here is, what should we be collecting, and how do you have this in a way that will be most useful?

Obviously it is a guess. If we knew the answer, we wouldn't have to be having the discussion. But I do think we need to think very carefully, so that we cannot lose 2 years while people are out building the tools that we need.

Dr. Falk: I think that is a really excellent comment. Do you have any last comment?

Dr. Colamarino: Along those lines, it was something I was thinking about from the earlier session. One of the take-home messages from yesterday's discussion was the crucial importance of phenotyping. That may indeed be a bottleneck, but not just phenotyping across all levels, but particularly getting at the biologics, what I like to call the wet stuff.

So I was pretty impressed by the fact that, although we didn't discuss it in any depth as a program, there is this Phenome Project that is working with the NIMH Intramural Program and with the M.I.N.D. Institute, and now I am hearing a lot more about the CADDRE as they are getting up and going, and there seems to be a question of biologics.

That ties into what Tom Insel was just saying, which is banking. I was really curious, listening to both of your ends, what are the efforts to coordinate those and make sure that the samples are collected in a somewhat analogous fashion, so that when these tools are ready to go, we can make use of them?

So is there cross talk?

Dr. Schendel: We certainly are talking, yes. We are talking, and we are banking. As I say, the collection protocols that we have developed in CADDRE hopefully will accommodate a variety of analytes to be tested. Since the phenome project is on the threshold, I think there is ample opportunity.

Dr. Swedo: And phenome grew out of CHARGE. So what Irva Hertz-Picciotto has been doing in her study is the basis for what we are doing. We have added a few things, but absolutely are using the same methodology and collection. So it won't matter whether they are from NIH intramural, M.I.N.D., or UC–Davis. They should all be comparable.

Dr. Spence: I think one thing to mention, though, is that with the genetics and the repository, that is a renewable sample because they are cell lines. So we have to think about the fact that these, once they are used, they are gone. I assume that we can't—I don't know of any biologic way of reproducing protein in from a blood spot.

So I think we have to think about that when we think about banking. It is not quite the same as the repository, where we can keep getting the cell lines and keep getting the cell lines.

Dr. Insel: And the genes don't change, so what you want to think about now is, are we missing something, is there some critical window that we need to have samples, and what those samples should be, that we will have wished we had collected 3 years from now.

Dr. Falk: We must now end this session, but first Alan Leshner has a few words.

Dr. Leshner: The next session is extremely important, and I would like you to think about what you are going to say before you say it. Drs. Martinez and Pessah have agreed to chair it. The idea here is to try to articulate briefly as many research gaps for as much feed-in toward the research agenda as we can get in. That is not to say we haven't done a lot, but in order to make sure that we have a full discussion, let's not have everybody say the same thing, just as a ground rule. If somebody makes a point, let's not beat it to death, but keep moving.

I want to repeat the comment that I made when we began. My own belief is that the research agenda needs to be as broad as possible and cover many items. From my own point of view, nothing should be off the table, but I would like to make sure that in the brief time we are going to have for that discussion, that we do in fact cover it.

Session VII
Future Research Directions—Discussion with Workshop Speakers

Dr. Leshner: Dr. Martinez and Dr. Pessah will moderate this session, but the purpose of it is to identify research tasks. So do one of you want to start?

Dr. Pessah: We were quite surprised that this would be the topic that we would have to moderate. What we tried to do, since it is a broad topic and we want to keep it somewhat logical in progression, was to identify the major problems that were brought up during the discussion. For example, how do we identify a study population, what should go into a study population, what are the comparison groups? So this necessarily involves epidemiologists, which need to identify a priori assessment tools, what sorts of medical record abstractions are needed, what sorts of questionnaires might be needed to get the most information that might in fact lead to an understanding of environmental exposures, even though questionnaires have their own limitations.

Then most importantly, to define a rational and doable strategy for biological sampling. One of the major hurdles is, how do you collect samples that are consistent not just across studies, but with-in a study? Samples collected at one period in time need to be comparable to samples collected 3 years later in the same study. So this is a very important issue.

I am going to let Fernando take us through the first part, where the discussion has led to the initiation of case-control studies. This is CHARGE. APP has a case-control design. We have also talked about MARBLES and the National Children's Study, which is more of a longitudinal assessment, and it is exposure based, and then a cross-sectional design. We haven't talked much about that sort of approach. Then, are there other approaches that we can engineer that would give us even more information or more accurate information?

Dr. Martinez: Just very briefly, because we discussed this yesterday, I think it would be very important for people when they come up with ideas for environmental studies to think about them in the framework of which type of design would be appropriate for the specific exposures they are interested in.

If the exposures can be relatively easily determined, I am going to give an example, duration of pregnancy could be an exposure, certainly the case-control approach is the best. You take a group of cases and a group of controls and determine the difference retrospectively of this easily understandable exposure, or relatively easily understandable.

For most exposures that are not easily ascertained retrospectively, the longitudinal approach is the best one. In other words, you start by defining the exposure and then you determine which is going to be the proportion of subjects with a certain exposure that is going to develop the condition or not.

The one that we have talked very little bit about is the cross-sectional design, which is particularly useful for suspected exposures that can be ascertained at the time in which you also ascertain the outcome.

I was talking about it before, and I give it only as an example, is Hispanic and non-Hispanic. That is something that could be easily done, quote-unquote easily, through a cross-sectional analysis. You take a group of subjects self-defined as Hispanic, or you can even do it by genetics, because there are now markers for ethnicity, and you determine by an objective measure what proportion of subjects have some type of autism-related disorder and what proportion do not within those particular groups.

There are other designs that we could talk about, but these are the three main ones. So it would be very important when we think about what we are proposing what way you think would be the best way to do it, and some of the epidemiologists here can help with that.

Dr. Pessah: So we are open for discussion.

Dr. Schendel: I just wanted to suggest that another opportunity might be occupational exposures, in populations with farm workers or other highly exposed families. They may be another group of individuals to do an exposure cohort type of model that can be used in this kind of analysis.

Dr. Fombonne: I'd like to come back to the issue of the natural experiments that we raised yesterday. One way to look at it is with the development of international studies in autism, to try to identify populations who have different traits, while surveying with the same methods.

I gave a preliminary communication last year. One population we have found so far where it seems that autism is somewhat absent. That is the population which is Inuit, living in Northern Quebec, where we have a population of 10,000, captive population in 14 villages which are constantly surveyed by our services, and we have found no cases whatsoever, despite very intensive efforts to find some.

So I just want to present that evidence as preliminary. I would like to confirm it, but I have some other parts of Canada with similar information, too. If so, that would be very interesting to follow through in terms of environmental exposure and also genetic backgrounds. In that case, the population is highly exposed to mercury and also highly exposed to PCBs and other kinds of neurotoxicants.

So it is just to give an example of opportunities which might develop. Other opportunities would be to look at plausible sources of environmental exposure. Here we have pesticides or mercury as an example, and to find situations where there is change over time, or geographical change in the rates of exposure, and try to relate that to risk of autism.

As I said this morning, there is an elegant study done in California where they use time-clustering and space-clustering methods to look at risk of autism in association with pesticide exposure. It is a very interesting finding that they have.

So these methods should be followed thoroughly, because they are already cheap and easy to implement, provided that they are guided by some kind of theoretical model.

I want to come back to the issue of the design. The case-control design to look at pregnancy is what we don't want to have. The prospective provides a very wonderful vehicle to look prospectively at a range of exposures. Using a fishing expedition, they are going to get samples and look at a range of possible things, and that is fine. But I think they are also limited in the fact that they started to survey pregnancy at age 18 weeks, which if you were to devise a model in terms of the timing of exposure regarding specific exposure for autism, you would try to look at what is happening before.

What we know from environmental exposure, like valproic acid, thalidomide exposure, were all in the first weeks of gestation, when it all led to autism. So I think there is a critical task for epidemiologists, to try to sample the possible exposure during the first weeks of gestation. The way to do that would be to revisit the first sibling studies, where you can have a high participation rate from families, you have an increased risk of outcome which is quite substantial, and you could look also at phenotypes, which are not only the full-blown autism, but also a variation on the same thing. I think that would be a way to enrich these studies by focusing the research on the biology.

Dr. Newschaffer: Just to follow up on what Eric said, the MARBLES study that started in California, we have in Maryland a pilot study already funded. The early network which has been mentioned, but not really explained in any depth here, will be funded as an autism center of excellence beginning in FY 2008, and it is exactly that design. It is recruiting moms who already have a child with autism at the start of a subsequent pregnancy. It will still be a challenge to get those very, very early samples, but that is what we are going to attempt, following them through pregnancy with repeated sampling, following the child with repeated sampling until 36 months.

So it is a high-risk cohort design. The pilot studies have already begun, and the full-blown study, which will attempt to recruit 1,200 pregnancies with a thousand kids followed up through 36 months of a 10-year period, funding will begin in 2008 and recruitment will begin in 2009. That is going to be done in four centers around the United States.

Dr. Schwartz: Let me ask a really simple question. Are the mental health folks comfortable with the way the phenotyping is being done in the environmentally driven studies, and are the environmental folks comfortable with the way the environmental phenotyping is being done in the mental health-driven studies?

Dr. Martinez: Who is going to speak for each of those? Who are the phenotypers?

Dr. Schwartz: What are the phenotypes?

Dr. Martinez: No, who are the phenotypers? My experience in CHARGE, we have had several individuals, experts in their own fields, come into the study with really no expectations. The ones who stand out are the molecular biologists who, when we discuss autism, they say there is no way that you will be able to detect a signature for autism using transcriptional analysis.

In fact, they were right. When we started to look at the transcriptional profiles of 25, 50 kids, comparing them to DD and to general population, there was no real clear trend, there was just noise.

Because we have Sallie Rogers and Sallie Ozonoff, and we have the very, very thorough psychometric analysis, the ADI, the ADOS, and other instruments, we could then subphenotype them into early onset, regression, spectrum versus full-blown autism. That really made a huge difference for the molecular biologists.

So I think in that respect, we could do better, but it clearly shows that if you have very well defined stratification, then it improves the molecular biology and the interpretation of the results.

Dr. Herbert: I would like to expand David Schwartz's question. There are the psychologists, the mental health people, there are the environmental people and there are the medical people, the people dealing with the medical illnesses of the children. The question of how those are going to be characterized also needs to be addressed. It is one of the areas that has been least well developed in the field.

I think that is important at the level of physical examination, and it is also important at the level of distinguishing between biomarkers of exposure and biomarkers of effect. I think it is important with biological sampling and with all of the questions that we are asking here, what is the impact on the individual patient, because that may have impact on interpreting the levels of exposure relating to the thresholds that may be impacted by the redox status or a variety of other things.

So I think that the middle needs to be addressed, and not just the behavioral and the environmental.

Dr. Schendel: I'd like to tag onto those two points. My first thought that I had before we got into this last issue of what is missing in the phenotyping fortified what Craig Newschaffer was saying and others. What we can do with these different types of study designs for either ongoing or existing studies, is augment the data collection for particular environmental characteristics and add to the data collection.

For instance, what we could do in the SEED study, and what is done in CHARGE, is to go back and get residential histories into data linkages with environmental datasets, identified at a very crude level, perhaps, but to some extent some sort of household exposures or residential neighborhood exposures, preconceptually, so being creative and getting additional environmental data that might be augmented for existing or ongoing studies.

You can't do all of this data collection on any one study, which feeds into my second point. When you are doing the phenotyping, feasibly there is a limit to what the individuals may be able to tolerate, both the cases and the comparison group that you are bringing forward.

So I think you do need to keep that in mind. Some studies may focus specifically on a phenotypic assessment to get as much detail as possible, while others may do sufficient data collection, for instance what CHARGE and CADDRE are doing, to do an adequate phenotyping, but perhaps without the mirror imaging and so on that are also beneficial.

So going back to the idea of coordination across studies, I think it would be important to come up with a core protocol, a core phenotype protocol, which then could be used to connect information across different studies in more creative analyses, and taking advantage of the wealth of data that is being collected across studies.

Dr. Insel: Some of that is actually in place. There is an attempt to have a core assessment battery. That is part of what NDARS tried to lay out, and to make it very easy by having all of the tools in one place available on a laptop, so you can take them and run with them.

I think David's question still needs to be drilled into a little bit, and maybe we can have a more thorough discussion of that. Before we get there, I am still stuck on that first line up here about biological sampling.

If I understood what Eric Fombonne was saying, in those rare cases where you have an environmental exposure that appears to be causative, as we sometimes say about certain genetic mutations that are so predominant that the environment doesn't count, and there are certain environmental exposures where the genetics don't count. So if we had a couple of those, just as we do in genetics, you would want to use those as your beachhead, to start off with.

Indeed, if it is the case, which I hadn't heard before in the discussion, that the ones we have, valproic acid, thalidomide, whatever else, which you could say would be hard-core discoveries, happen very, very early, it seems to me that ought to change the nature of the discussion about biological sampling.

If we had had this meeting 2 years ago, we would probably be talking about what samples should be collected in the second year of life, or maybe toward the end of the first year of life. Most of our discussions have been about prenatal sampling, meaning second, third trimester. But if it really is first trimester, then we need a very different kind of discussion about sampling, as you are suggesting.

Am I correctly understanding that proposal?

Dr. Fombonne: This model is for the fourth or fifth week of exposure. All the data that we know about valproic acid and thalidomide suggest exposure during a very narrow time window, which is around the 24th, 25th days of gestation. It is also consistent with neuropathology findings, like Margaret Bowman and Thomas Kemper have also related that back to the first trimester of the pregnancy, most certainly. So there is converging evidence that the first trimester is probably a critical time point.

Dr. Pessah: I'd like to speak to this as a toxicologist, because those models certainly are interesting models of autism, but they don't really portray the full spectrum of autism.

One could imagine that an early hit could devastate the developing system in such a way that one of the endpoints is going to be autism, but there are also significantly other teratogenic effects that these models produce. We know that we can have autism without teratogenesis.

So although they serve as a valuable model, I would like to point out, for example, to the valproate model, where it has been linked to *HOX-1*, and now there is some linkage to the *HOX-1* gene, but it is a downregulation of the *HOX-1* product, whereas with valproate, I know there are some data out there now where it increases in response to valproate.

So you have to be very careful with limiting the spectrum to a very early point in development. It may be relevant, but it doesn't fully explain the full spectrum.

Dr. Susser: This is something that I have done a lot of work in. I agree with Eric Fombonne, in the sense that we should get samples as early as possible in pregnancy. That doesn't mean that only very early is important, but we have suggestive evidence that very early is one of the important time periods. So we should be aiming for that.

There are indirect ways to get that evidence prior to the National Child Study, where it is actually built into it, but that will be quite a long time before that actually happens. One indirect way is to use the first prenatal visit as an indirect proxy for what is going on in early pregnancy. It is not perfect, but it is something that we can do.

We have, for example, information at 17 weeks, what the mother was taking at 5 weeks, and you have information from blood samples from the mothers about things that may not change that much between 5 and 18 weeks. It is indirect, but it is still much better than anything else we have.

The second thing we can do is something that Dr. Fombonne also alluded to, which is similar to what we did with the famine study, which is seek populations who have been exposed early in pregnancy to some of the toxicants that we are concerned about. There are such studies, but I think Dr. Schwartz's question is relevant here, because those kind of populations, I don't believe have yet been used to study autism and those kind of outcomes. So that is another way that one could solve the problem.

I would agree with you that the focus shouldn't be exclusively early pregnancy, but that should be an important focus.

Dr. Beaudet: I think that a biological sample should be collected as early as possible from the father, if you are going to collect biological samples. We know things about advanced paternal age, and there is plenty of potential for other paternally originated abnormalities.

Dr. Martinez: Should we go to the next point, because endophenotype seems to be very important?

Dr. Pessah: Yes. One of the major points that was brought up is endophenotypes and better stratifying autism, if anybody would like to comment.

The idea would be that you have to have communication between the epidemiological design and how you are going to substratify the cases based on very detailed diagnoses and maybe medical exams as well.

One of the things that I think is a gap is, we haven't paid a huge amount of attention to comorbidities, including what seizure disorder might tell us about substratification, or cardiovascular problems. Some of the genes are expressed both in terms of the cardiovascular system.

If you want to go out there, nobody ever collects data on family history of different kinds of cancer. Given that PTEN and Met and some of the other genes we have heard about are very, very highly involved in human carcinogenesis, is there any link there?

Ms. Bono: We talked yesterday, one of the ideas was to data mine some of the practices. I know that every time I have taken my son in, they have asked, heart disease in the family, cancer in the family? They ask all sorts of autoimmune disorders in the family.

So again from yesterday, I think we need a formal mechanism where we can send investigators into these practices and data mine what is there from the kids. That leads into the recovery study as well, what can we learn from the recovery period.

Dr. Martinez: But in that case it would be extremely useful to have some kind of instrument.

Ms. Bono: Right, a formal mechanism.

Dr. Martinez: That has to be not too complicated, either, because many of these practices are extremely busy practices. I am telling you my experience in asthma. We have done this in our work in longitudinal studies, and the best way is for those who are experts in the field—this is something that David Schwartz was asking for—to define a minimal set of ascertainments that need to be done in a very large number of potential cases. I don't get the feeling, I may be completely wrong, that such a thing exists. It would be very good for some consensus to be created as to the development of such kinds of tools.

Ms. Bono: I agree with that, and to not count on the practitioners to be able to do it. They are just too busy.

Dr. Martinez: Right, but you could develop even some automated systems that can be accessed through the Web, in which you just fill out one form. That is what we have in asthma. It is an idea that could be developed for autism, too.

I know it is a complicated disease, but believe me, asthma is also complicated.

Dr. Swedo: Is Paul Law here this time? There you are, Paul. Do you want to take the mike and respond to that for IAN?

Participant: The IAN project is an online environment where families are invited to come and provide information that they can provide accurately based on their own experiences with autism in general, everything from the school system to their diagnostic workup, the day they were diagnosed, and so on and so forth.

One of the more valuable parts of what we are collecting is the treatments that parents say that they are on. It seems like it is pretty accurate—we try to do a lot of things to assess the quality of the data that are coming in, but I think the treatments that families have their kids on, they know this pretty well and are able to comment on it pretty well.

It is very interesting. There is clearly a different practice. There are different camps in terms of the way families are being managed.

I think that capturing data from the clinical environment as well would be a really good thing. It is going to give you higher quality data than what we are able to capture on IAN. The nice thing about IAN is, you create a website, the next day it is available to everybody, and you have some very dedicated data collectors coming in by the four thousands to provide information. A lot of noise issues can be dealt with, with this enormous sample size. Literally we have had over 3,000 people join in 2 weeks.

Dr. Spence: Can you describe what the general questionnaires are? Are you doing a medical history? You talked about the treatments, but in terms of thinking about environment, are you doing an exposures questionnaire, that kind of thing?

Participant: Yes, I think there are opportunities there as well, but you always need to be cognizant of the limitations. The content in IAN right now is wafer thin, because we wanted to make sure we didn't overburden families. It was mainly designed around making sure that we had enough information to help researchers be able to identify subjects. We wanted to at least be able to do that very well, create a very efficient matching system where researchers can identify subjects in the areas. We can e-mail them, and we can try to illuminate this thing that is slowing down autism research.

It seems like every researcher I talk to has problems getting subjects. If there is somebody who has solved that, let me know. But there is a huge amount of potential. I get 10 e-mails a day from mothers wanting a pregnancy history to be added that is extensive, or immunization histories to be added.

In general, families are very interested in providing more data. I think for certain types of studies this could be at least good preliminary data to build off of. It is certainly not going to compete with nice, sound epidemiological population-based studies, but it is one of these things that we can do really quickly. Like the terbutaline thing, we can have that done in a couple of months with a sample size of 4,000. We can really move things along pretty quickly for certain types of questions.

Dr. Martinez: I think you are referring to family-based methods. That's fine, I think it is very useful. The caregiver base would also be important and perhaps could have a different type of role.

Since I am involved in that field, I have to say, for example, that in the field of cystic fibrosis, there are very interesting developments. Some of the things that we have found out about cystic fibrosis were based on a large epidemiologic study which is caregiver based. For example, the importance of diabetes in the prognosis of cystic fibrosis was ascertained through these kind of methods.

So the development of tools for standardized acquisition of information from caregivers would be a very interesting development.

Dr. Law: I'll make one comment about that, and I'll get down. One of my degrees is in medical informatics. If you look at private practices or community-based practices, less than 10 percent or so have any kind of electronic data capture happening in their practices. So if we are going to try to capture all these clinical data, we are going to have to act very strategically and try to solve the problem together. It probably needs to be a Web-based system, and we need to think about electronic health records, creative ways of addressing that issue and capturing it.

Dr. Leshner: Can I just make a suggestion? I am a little bit nervous that we are spending way too much time relative to what we have designing specific studies. So if we could identify the content gaps, I think that would be very helpful.

Dr. Pessah: Just content gaps. Anybody else want to comment?

Dr. Beaudet: Just a brief point. I think it is really important for those of us who are studying autism at different levels, in my case genomics, to have the appropriate controls available.

So far, although I can use autism-unaffected sibs from families with autism, I don't consider them to be the best controls. There is no national repository I can go to to get children in that age group. I can get adults, but I want to compare against my children controls where there is no autism in the family.

Dr. Needham: One gap that we have talked about somewhat as far as biological samples go is what kind of chemicals to try to measure. I was just wondering, if oxidative stress is a potential pathway, is there an in vitro or some kind of test that we can use for various chemicals to test if those chemicals have the capacity to cause oxidative stress, and use those perhaps as biomarkers of exposure?

Dr. Noble: Sure. That is the kind of work that I talked about yesterday. You can run as many chemicals as we can grow cells. We have sensitivities, mercury, four parts per billion, lead, low ranges, ethanol, 17 millimolar, low end of the clinical range, everything we look at.

If you look at progenitor cells taken from the right time in the developing nervous system, you have these extraordinary sensitivities and the capacity for high throughput in multiple endpoints.

Dr. Needham: Are you saying that any chemical can cause oxidative stress?

Dr. Noble: No, I'm saying that any chemical that causes oxidative stress will have a set of well-defined endpoints that can be analyzed.

Dr. Needham: So can you test a large array of chemicals, say 1,000 chemicals, and say that 500 of those cause oxidative stress and 500 do not?

Dr. Noble: Yes, absolutely. We do that all the time for various purposes.

Dr. Needham: I'd like to see that.

Dr. Pessah: I just want to add one gap here that is relevant before I go on. That is the gap which deals with the cost of extracting quality information from medical records. We need to put more money into that. These studies regardless of design really rely on the data that you extract, and there is no easy way to do it. It is all done by hand at this point.

So we are going to go on. Exposure assessment. There was a huge, huge issue about what we measure, what is our sensitivity, and how do we integrate large datasets on exposure assessment. This would involve exposure experts, analytical chemists, engineers who are developing new technologies, and of course through the new NIH genome environment initiative. We need to get new technologies in place, and we need to do it as soon as possible.

One of the things that I think we need to identify gaps is in terms of prioritizing which chemicals and which physical agents that are accurate or predictive of the environment in which the study population is living or has lived in. This would be different, I would imagine, for agricultural communities such as those in the Central Valley of California versus urban communities, let's say in New York City.

We want to be able to optimize environmental samples. What does that mean? How do we stabilize across studies so that these environmental samples have the same information that we haven't changed in the process of either collection or storage of the samples?

There are no real guidelines at this point. I think we need to set up guidelines. I think there are some guidelines maybe through the CDC, but are those the guidelines that everyone uses for the studies that have been mentioned that are in progress or just beginning.

Implementations and development of biosensors. I think this was a perfect lead-in for this discussion, because I think the technology has progressed to the point where very soon we will be able to make thousands of measurements simultaneously. Could that impact the study design, the fact that we would be able to measure everything, or everything we think at that point in time is important, rather than having to make choices? So I want to open that up.

Dr. Noble: One thing that I don't see on there, and I want to put this in perspective. I live in multiple medical communities. We work on spinal cord injury, we work on cancer, we work with developmental disabilities. In all of these communities that I work in, I have to say that I have never heard patient stories like the excreter phenotype story, like the recovery stories that I am learning from talking to patients here.

If I hear those stories once or twice, I have the standard reaction of a scientist of being an anecdote. But I am hearing too many similarities.

This issue with the excreter phenotype, I have to say I find it both fascinating and worrisome in this trial design. If this idea is right, that at least some subset of these kids are accumulating body burdens because they cannot excrete, they will never be picked up with that design.

So I think there needs to be a part of this where somehow we are getting improved access to the ongoing patient experiments that are going to go on, no matter what we decide in this room. They are going to go on, they have to go on, because that is what the parents need to do. Somehow we have to be willing to solve this scientific question of, are there kids who when they are treated with chelation therapy or antioxidants or whatever is being done, are they pouring out heavy metals? As a scientist, I would really like to know that.

Dr. Insel: Why couldn't you do that in a challenge approach?

Dr. Noble: Right, and that would be incorporated here. That is what chelation is, it is a challenge approach, right?

Ms. Bono: I think what you are saying, though, is the difficulty of measuring when you are pulling these samples. They are not done provocatively. Trying to get the chelation study off the ground, as we are all hoping to, but pulling the data samples without a provocative agent may not produce any information.

Dr. Noble: That is what I am worried about, from what I am hearing.

Dr. Swedo: I think you speak to a bigger issue, Mark Noble, and I hope we can let Isaac Pessah finish this. I see this as being one area, that is, are there population differences for children who end up with autism versus those who don't end up with autism, and the increased evaluation one can do to those groups. But I still am not satisfied that the population-based approach is the only way to go.

I again go back to the example of leukemia and diabetes and other things, where you start with the patients, and you start looking for differences within those kids, so you would use a clinic-based approach as well.

Dr. Martinez: That is why at the very beginning we said that there are several different approaches. Many of these are longitudinal, the ones we are talking about now, but the case-control approach has its advantages.

Dr. Swedo: I don't even know that it would necessarily have to be a case-control.

Dr. Martinez: For example, it could be that cases are the ones that are improving and controls are the ones that are not improving. These are mainly epidemiological. We can add to this that is much more a case control type than the type we have talked about.

Ms. Redwood: I just had a real quick comment. One of the things that we have been focusing on is the heterogeneity of autism. My concern when I hear about these chelation studies is that we are not doing biomarkers ahead of time to identify the children who actually have a heavy metal burden.

There is a test available, it is a urinary porphyrin test. I would say that we first try to delineate the heterogeneity of autism and then provide targeted treatments. If you take every child with autism and enroll them in a chelation study without first identifying the children that have body burdens of metals, then the ones who do respond, it is not going to be significant because the other ones that didn't have a body burden of metals to begin with.

Dr. Martinez: I think we should not discuss, as Dr. Leshner has asked, the details of different studies. We have a general agreement that some type of therapeutic approach needs to be tested, and I think that is what we should say. Why don't you please continue, because otherwise we won't finish.

Dr. Hertz-Picciotto: Can we go back to that last? I did want to say something about exposure assessment.

Dr. Martinez: Yes, go ahead.

Dr. Hertz-Picciotto: There are a couple of things that I wanted to just mention. First of all, in terms of the chemicals and the prioritization, there is a whole class of compounds, and it has been alluded to in a couple of slides that no one has started looking at, at all, as far as I know. We haven't started, and I don't know if CDC is planning to, in the cosmetics.

Now, cosmetics are primarily used by women, and they don't usually just stop at the point when they become pregnant. So that is a whole area that I think does need to be addressed in this field.

The other aspect, in terms of how we have been prioritizing, I think most people have been thinking in terms of direct neurotoxins. I'm not sure who it was yesterday who gave a list of the ways we might think about direct toxins as well as factors that indirectly affect the development of the central nervous system.

In particular the hypothesis around inflammation has come up in the literature in a number of different contexts. The work from our center seems to suggest immunologic dysregulation going on. So I think we need to think in terms of toxins to the immune system as being suspect agents that ought to be on the list.

Then in addition to the issue of biological sampling, I just wanted to raise the issue of general environmental sampling. Diana Schendel referred to existing databases that are out there for exposure assessment in general, but we also might be thinking in terms of this whole issue of sensors in the home. Indoor air is a lot more contaminated than outdoor air. Most of what EPA does is outdoor air monitoring. So a lot of the chemicals that are generated by our textiles and other home products that we use for cleaning agents and so forth have a great deal of toxicity and ought to be on the list.

Dr. Pessah: We should probably move on.

Dr. Hertz-Picciotto: But anyway, household sampling, not just biological.

Dr. Insel: Same slide, last line, where you say analytical methods, I would include some way of using the genome as a sensor, whether it is epigenetic changes or DNA damage.

Dr. Pessah: Yes, that comes right here.

Mr. Blaxill: Do you guys have a guide to the framework, so that we can be trying to absorb it while we are talking about it?

Dr. Pessah: I'm sorry?

Mr. Blaxill: The framework is being progressively revealed, and it is hard to—in your head without knowing what the framework is.

Dr. Pessah: I should have had it all planned out ahead of time in terms of a diagram.

Dr. Hertz-Picciotto: We should have let him finish presenting before we started picking it apart.

Dr. Martinez: Everybody wants to say something.

Dr. Schendel: Just real quick about the exposure assessment. One gap is the understanding of the exposure response in pregnancy specifically of the woman who is exposed to a particular compound or chemical, how the body is processing that exposure. Larry Needham had in his slides how it is being shunted to different compartments in the body. Those are typically done on nonpregnant subjects, I would imagine. I think the pregnant state would be an important gap to fill.

Dr. Martinez: Why don't you finish the whole thing fast?

Dr. Pessah: Okay. Clearly to try to understand genetic differences between some groups of autistic children and the general population, we can now do this much more efficiently with the technology that has matured. Five hundred thousand single-nucleotide polymorphism chips are now available. If you want to look at particular sections of chromosomes like long-arm chromosome 10 or short-arm chromosome 7, you can do so with a nimble gene array, which allows you to get very high resolution information about low-copy repeats and so forth.

Epigenetic information is a bit more arduous and time-consuming, but one can see if there are major differences in methylation in any one of these designs. Metabolomic, proteomic, and lipomics are the -omics profiling. There has been this amazing advance in how many analytes we can measure at once.

I know David in Bruce Hammock's lab can give you every single intermediate in the methionine pathway all in one shot. The question is, are the samples stored properly so you are not just looking at variation, which goes back to sample quality.

Toxicologists and cell biologists, we have real difficulty collecting brain biopsies for cellular studies, mechanistic studies. It is just never going to happen in this disorder. However, I want to point out that many of the candidate genes that have been reported to date are expressed in the immune system. So we need to focus a bit more on the immune system as a biomarker or a target in autism.

Here I have given you some examples, where one could immune profile from primary cells. I think that cell lines, Epstein-Barr transformed B lymphocytes are very useful, but they have their limitations. So primary cells give us more specific information, and will probably be more predictive, especially when you are trying to subphenotype.

Intracellular cytokines can now be measured, so you can measure cell activation parameters for intact cells. You can do antigen recall and see if there are problems with antigen recall, and you can also challenge with environmental agents, endotoxins being one, but you can also, once you have identified candidate xenobiotics that are of high probability, you can then apply them to a cell-based system to see how the signaling response of immune cells differs.

And of course, one can look for autoantibodies to brain proteins, which sometimes we forget. If the mother has autoantibodies in circulation during gestation, the IgGs do cross the placental barrier and can have an effect. It is a genetic problem which is transformed environmentally within the developing fetus.

One could also use immune cells, primary immune cells, to do a detailed analysis of signaling cascades. Here, we don't have to identify which kids have the mutations ahead of time, because these signaling cascades converge on common denominators.

I have just named a few of them here, and we can talk about them. Met goes through Grb2, Gsb1; it impacts pI3 kinase, which then can influence all of the targets of pI3 kinase that influence cell growth and proliferation. Cav1.2, it is part of a large complex that regulates local calcium signals in cells, which is incredibly important for just about every cell process we can think of. It is also regulated by pI3 kinase.

So we can look for commonalities in cell signaling that is aberrant in autism cases without knowing the mutations ahead of time.

Animal models. I think there has been some negative press here about, you can't do a single-gene mutation in a mouse because it won't be representative of the condition. But I would like to bring up for discussion for transgene X knock-in and knock-out mice are in fact useful, especially if you can time when the mutation occurs, and look at specific endpoints in signaling abnormalities that may be relevant to what you discovered in the cell models that you did get from autistic kids out at the periphery. Here you can look at the immune system and the nervous system.

Finally, one can humanize the mice in a very different way. If there are roles of autoantibodies for autism susceptibility risk, one could humanize the mice by exposing them to the human IgGs and see if they develop neurodevelopmental defects.

Other models include the PKU model, which is not an early model. Poly-IC, I think we have talked about the early models, poly-IC models that Paul Patterson has initiated work on, and oxidative stress models. We can test those in animals thoroughly.

Data validation, integration, and modeling are absolutely essential if you are going to put an integrated effort together.

Finally, and this leads to the next round of talks this afternoon, is the randomized controlled trials of novel therapeutics and innovative intervention. Chelation is one that has been discussed quite a bit here, but also what about antioxidants, vitamin supplementation, what about DHA, some of the biological markers that are oxidative stress involve lipid metabolism. How can we bring these into innovative clinical trials? Then finally, combinatorial therapies.

Dr. Martinez: Opinions?

Dr. Noble: One of the things that I did not hear, and I understand its complexity, is the understanding of toxicant synergies. Having recently felt that our mechanistic analysis allowed us to explore this, I was horrified by how little knowledge we have in the literature about this, yet here we are very interested in multiple exposure paradigms.

So I think that is a biological question that I think is addressable, and I think it is critical to understand. From what we are learning, it is also going to teach us a lot about the underlying mechanisms of susceptibility.

From what Gil said in regard to sample collection, one of the notes that I took was the critical aspects of the timing of sample collection, and whether the individual has had a meal beforehand. If you are going to do metabolic profiling, you have got to take into account that metabolic profiles change, and circadian rhythms. They change according to what we need to change according to when we ate. So I think that has to be built into the trial design.

One of the things that we have not discussed yet that I would love to hear information on, if not in this setting, then eventually, in the pediatric neurology community, where there are so many kids that need some kind of treatment, one of the points that I have been making repeatedly in the meetings that we have is, tell me what I am supposed to repair.

I am a stem cell biologist. In spinal cord injury, I know what I am supposed to repair. I can design experiments. I know what I am going for. So I am a total newcomer to this area.

Before I came to this meeting I read widely. One of the papers that many of you know about is this mirror neuron study from Rama Chandra's lab, which is so intriguing because of the role we think mirror neurons have in enabling recognition of the other as self in a study. That needs to be repeated, it needs to be extended, but it raises the possibility that kids with autism don't have mirror neurons.

Well, if I know that as a guy who is trying to do repair, it helps me to think about it. I need to know a lot more like that. It is part of why larger studies are so important.

The last note I have here I have mentioned a couple of times, and I just want to come back and emphasize. What I am hearing from the patients in this setting is unlike any other setting I have ever been in. This is one of—in more than 25 years of doing clinically related neurological research, this is the first meeting where I am coming away from it saying that the patient populations are doing experiments that we need to pay attention to.

They are telling us a lot that concerns me about trial designs. I have been talking to a number of the parents and saying you do the chelation therapy, and without doing secretin and it doesn't work so well. I think we have to draw that information out.

Dr. Schwartz: I just had one more item to add on the list, which is data sharing and general access to data, both the genetic and the phenotyping data. I think it is something that we should consider.

Dr. Martinez: That is a crucial point in all areas of research, but I have the feeling that particularly in this area it is crucial.

Dr. Leshner: Does this community have any implied at least policies around data sharing and database sharing?

Dr. Martinez: That used to be a difficult question.

Dr. Insel: It depends what you mean by the community.

Dr. Leshner: Or the database. In the ACEs, that is, the autism centers of excellence, it is a requirement that all the data goes into NDAR. But as you are hearing, there are many different communities which are not at this point integrated together. As David Schwartz suggests, that is an opportunity.

Dr. Martinez: And it is also a challenge. This seems to be a need. I will give again the example of cystic fibrosis. There is a central foundation that directs the availability of data, and it has been extraordinarily useful, extraordinarily useful for us in the CF community. In any event, just as an example.

One of the things that could come out of here is precisely what David Schwartz has just asked for, which is mechanisms of data sharing. It would be crucial.

Dr. Beaudet: The Autism Treatment Network, their entire vision was to emulate the CF center concept, both from research and care and so on. Although I don't know what its latest merger implications are.

Dr. Colamarino: I was just going to say that, at least from the perspective of what Cure Autism Now (now merging with Autism Speaks) has been doing over the last 10 years, by building a database with the genetic information, it is all open access; it is all forced sharing. The problem is integrating it on a larger level, which is supposed to be attempted by NDAR, with respect to the medical records. It was based on CF. And there is the Autism Treatment Network, but they are just getting off the ground, which is partially the reason they are not there yet.

The idea was to create standardized medical assessments. All that information would be put into a communal database that could be mined by whomever. But they are just getting off the ground, which is partially the reason they are not.

Dr. Spence: I think the opportunity for sharing is great, because people have been recreating the wheel a lot. This is something that there is a great medical history that the ATN is using based on the AGRE history that we have developed for NIMH. But then there is also CHARGE, and the ACE centers are going to be using the CHARGE medical history.

Dr. Colamarino: We use that one.

Dr. Spence: So there is some overlap. I think that is something we can work on as a community. But I think providing those resources to other groups, for instance, that we can then—what I really wanted to say was, would it be helpful for us to sit for 5 minutes and have everybody mention one study that is ongoing that we know of that is collecting data, and then get in the room representatives from those studies so that we can talk about this sharing issue.

Dr. Martinez: I don't know. Dr. Leshner, I think one thing that could be done, I don't know if it can be organized, but some kind of a questionnaire or some document would need to come from everybody who has those type of studies, that could be given to the organizing committee. I don't think we have time today to go through that.

Mr. Blaxill: I think you were asking us to think about gaps, so I have two categories of gaps. But first I want to endorse what Tom Insel said, which is this notion of the integration. The communities are not well integrated, and I think that is a problem.

I think actually the environment is one of the reasons—the relative lack of effort on the environmental side is one of the reasons there is lack of integration in this. It would be helpful if this discussion leads to more integration. But there ought to be some progress on that front.

In terms of gaps, I would list two categories, and they are very different. On the one hand, I am thinking more about yesterday. There were a bunch of discussions about mechanisms, all of which raised questions about biomarkers. Dr. Insel, you asked the question about biomarkers. Biomarkers presume a model of what you are measuring and why you are concerned about it. I think there is a critical gap.

There is a real question about what is going on, what brings autism together, what makes it a coherent disease entity, if indeed it is one. I think there is a bunch of problems in that. There are hypotheses in different levels of scale, at the level of the topology of the network of the brain to tissue and pervasive tissue growth, or at the molecular level with signaling and redox balance and that sort of thing.

I think that is a gap. As a consumer of the science, I would love to have a clear idea of what my child has, and why she is like Ms. Bernard's child and why she is different, some kind of mechanism that clarifies some of the complexity and gets to final common pathways in a meaningful way.

I think success in that kind of work would—we would end up renaming autism. We would have another name for it that would be much more specific as to what the mechanism is. That would be a satisfying outcome and some distance from that.

The integration of some of the different systems, the interdisciplinary teaming, multidisciplinary teaming, I think is critical to make progress there. I think you all are doing more of that, which I think is great.

The other gap that is a different category is, I hear the enthusiasm among the professional epidemiologists for prospective studies. I know that from a methodologic standpoint you can control it, you can do it right, you can do it the way you want to. I think there is a gap in terms of high-quality retrospective studies. Recognizing that there is imperfect information, natural experiments; I'd love to look at the Inuit population. You could learn a lot from that. You could also triage some things about genetic versus environment. Carefully done, that could be helpful. The Amish community that comes up. They all have a lifestyle. Maybe we all need to follow the Eskimo lifestyle. The Amish have a lifestyle. Is that a risk factor or a protective factor, or are there other things going on genetically? Are there protective Amish genes? I don't know. There are reports that the Amish have low rates of autism. That would be interesting.

Going back to the original cases first discovered by Dr. Leo Kanner, Martha Herbert mentioned that all of the 11 Kanner cases had somatic symptoms. A lot of them lived in the same place. Was that a signal or just an accident of the diagnostic pathway? I think there is something to learn from that.

I think we need to face up to the question of trend. That raises retrospective problems, but I think it is critically important strategically. I obviously have a strong position on that, I have tried to articulate that.

I think we need to put that behind us. My personal bias is, if you have a 10-fold increase and there is a hypothesis that it is artifactual, the burden of proof is on the person making the hypothesis. We ought to test those theories and get them behind us, because we have been delayed for a long time by the confusion, and maybe a reassuring sense that everything is fine and there is not a problem. I think it is a strategic question when you talk about the environment.

That can only be done retrospectively. There are going to be imperfections when you do that. It is not as clean as a prospective study, but there are lots of natural experiments there and information that you can draw on.

Dr. Herbert: One more retrospective study that could be done is documenting retrospectively claims of parents that their child has recovered.

There is a registry online, and the campaign is starting to get more people to register their kids. In the interest of disclosure, I have an IRB to do this, but if we could go back and find kids who did have reliable diagnosis and follow them up and reliably assess them, that would have very important implications for the neurobiological underpinnings of our hypotheses about what it is that we need to measure.

Dr. Martinez: As a trained epidemiologist, I have to say that this goes back to what we were saying before. In order to be able to know what is characteristic of the children who get better, you need to know what is characteristic of those who don't get better. So you need good information from a large set of patients.

Once again, the availability of tools that will allow you to assess that, so that you have it at one point, and then 10 years later you can assess retrospectively, based on how kids got along the way, who had a certain characteristic at the beginning and who didn't, would be crucial.

Dr. Noble: I agree with that, but it is important to recognize in the context of neurological disorders, how unusual it is to have these kinds of improvements. You don't see this with spinal cord injury. You don't see this with cerebral palsy.

As a scientist I would like to have what you have, but also as a scientist if I had what Martha Herbert is trying to collect, I would be in there trying to get interesting data out of it.

Dr. Swedo: That speaks to the point that I was going to make. I think there is a huge gap in our definitions. Definitions of what is autism, we heard that in Mark Noble's comments, but also earlier when we were trying to talk about these epidemiologic studies. We continue to change the definition of autism spectrum disorders. We need to get back to more continuous variables, because our discrete categories are not working very well.

The second issue is definition of recovered. I was speaking with some of the practitioners and folks in the audience yesterday; recovered to me is very different than recovered to some other folks. I would want to know that we have a standard definition of response and nonresponse.

Dr. Martinez: This word is coming back very often: standard. The other one that comes is collaboration between groups, which is the other side of the coin of standard.

Ms. Bernard: I think there is a gap in tracking individuals with autism as they age. Some of these people become more severe and sometimes they get less severe, even if they don't fully recover, and there could be an environmental component to that. Do they have ongoing exposures? Do some of the differences in degree of severity that we see or onset of comorbid conditions, does it have to do with something in their environment? If we could get rid of later exposures, maybe it didn't trigger the autism, but it changes the course of disease. I think we need to look at that.

In addition in terms of gaps, for the epidemiology I think it is very important that we get a handle on subtypes, and that we don't just report rates of autism spectrum disorder as 1 in 150, or 1 in 100 for New Jersey, but that we actually determine whether that is Asperger's or PDD-NOS. Some of our current studies are not able to answer that with their methodology.

There was one other. I can't remember what it is. I'll think of it.

Dr. Schwartz: I just wanted to get back to a point that Tom Insel raised before which is related to acquisition that could be used 5 years from now, 3, 5 years from now. It relates to the new measures that may become available.

The question is, should we think critically about developing a biobank for samples across these different epidemiologic studies that could be centrally located, available to investigators, very similar to what NIMH has set up with the genetic resources.

Dr. Newschaffer: Another issue in the class of biomarkers, and I don't think this is a repeat, biomarkers of exposure susceptibility. We have talked about efforts in genetic markers of exposure susceptibility, but I wonder if there are gaps in the area, and the toxicologists can help me, in terms of phenotypic markers, other biomarkers of exposure susceptibility.

As an epidemiologist, if I am going to study exposure effects, I would really like to know who is susceptible and who is not susceptible to those exposures. I get a sense that there are some gaps there.

Dr. Susser: Just a very quick point. I am just afraid that we might miss the forest for the trees. I agree with all of the trees that we have talked about, but I feel that we do need to tackle the question of the time trend and what has happened, whether it has been stable or whether it has been increasing over the period in question. It is not impossible to do. All the things we are talking about are difficult to do, but that is one that I think we ought to take on in terms of gaps. We haven't done it yet.

Dr. Martinez: Any other thoughts?

Ms. Bono: I agree with that about the time trend data. The CDC said this morning that it was going to be difficult, but we are dealing with something that is very important. We have to find out what changed in the environment and when, and try to figure it out.

As I made notes throughout the last couple of days, I have identified several gaps. I'm not quite sure, Isaac, if it is in your things that you have mentioned—methylation and looking more into that. That is a fruitful area, so we need to make sure that those studies are ongoing.

We didn't have up on the chart, although I know a lot of people have talked about it, the toxokinetics of mercury, the transport mechanisms. We need a better chelator for the blood–brain barrier. So I would love to see something like that. We need to understand the transport mechanisms in the body and how it gets into the brain, how it can possibly come back out.

With methylation comes detox support. I saw you had an antioxidant protocol, so perhaps that is methylation/detox in some of that.

We identified this morning through the exposures we cannot identify, or we haven't yet, the nonvaccinated kids. I think that is an important gap that we aren't testing. When we see the large population studies, so much of the population is vaccinated. So we need to try and find those kids. Mark Blaxill mentioned the Amish, but there are also several medical practices across the country. There are education records, the waivers. We can go in and find those kids and be able to see what is going on there.

Lastly, gastrointestinal studies. So much is going on in the guts of these kids. There are failing enzyme systems, there is bacterial dysbiosis, there are gut pathogens and virus, fungus. Certain phenotypes have this, and can remarkably get better if these things are addressed. So I would like to see a gut study.

Dr. Wilcox: I haven't heard any discussion about what happens to these kids as adults. It seems to me that the natural course of the disease into adulthood ought to be information that would help us understand the etiology as well as progression.

Mr. Blaxill: The Leo Kanner natural experiment would be—they are 70 years old, the first 11 cases. Some of them are still alive. So that is one way to use that kind of information. I think, Sallie Bernard, you raised that point as well.

Dr. Martinez: We are almost at the time, right? I knew you were going to tell me that.

Dr. Leshner: I have decided that during the discussion session later, we are going to go around this table, everybody is getting one minute, literally one minute to identify a residual gap, not design the experiment, not design the instrument, content, what is the gap. You may pass, you may not pass, you can do it however you want, but you get one minute. I am really nervous that we are losing something, that we are not capturing something. So that will be the first exercise.

Dr. Martinez: I have the opposite feeling, that we have too many things on the table, but that's okay.

Dr. Leshner: For me you can't have too many things on the table. I think this is sufficiently understudied. Anyway, this was a good session. You can have one more minute.

Dr. Martinez: Let me try to summarize as an outsider the areas that I feel you guys and ladies have talked about.

I think there are issues that can be measured in individuals with the disease. I think that is the first area. I think we all have said that needs to be measured from the time of conception—of course, we cannot measure them in the individual him- or herself, but at least in the mother. Eric Fombonne has stressed many times the need for this to be from the time of conception, because things can happen in the very first month that are crucial.

These things that can be measured are phenotypic or clinical, genotypic, genomic, and metabolomic and all of the other things that we have talked about, and could be predictors of disease and consequence of disease. So those two areas need to be distinguished, I think.

The second is what can we measure in the environment? Everybody has said a lot about that. Since everybody is going to say something in one minute, it would be good that you would try to tell us which of these areas that I am trying to identify are interesting, or even other areas.

I think that these two measures—because in the end, science is measuring, these two areas which we can measure, I think we will need to concentrate ourselves in trying to understand.

The third area that has been stressed, which is potential models, are models relevant, which models? These models could be in silico models. Everybody has said the importance of bioinformatics, but also it would be important to know animal models, are they relevant, which animal models, which are you interested in?

Mark was talking about mechanisms. Unfortunately in humans it is tough to study mechanisms, but that is what animal models are good for. So perhaps that could be something that could be added to the discussion.

So those three I think are the main areas that people need to talk about. Then this can be transformed into specific studies. But I don't think it is necessary for us to discuss here the details of each specific study.

Dr. Leshner: That was really well summarized. Thank you.

Session VIII
Public–Private Partnerships

Dr. Leshner: This afternoon we have two parts to our discussions.

The purpose of the first session of the afternoon is to talk about public–private partnerships and to try and see whether there are things we can do that we cannot do independently, to try to see if there are unique opportunities here. In a minute I am going to ask each of the people listed on the agenda to make 5 minutes of comments. They can do that from their seats, but the light will be blinking nonetheless. Then we will have some discussion of that topic, but then I would like to reserve the last 45 minutes or so for a residual discussion of the gap issues and things like that, to make sure that we have captured everybody's major thoughts. We don't want your minor thoughts, but your major thoughts.

Let me just say that I think this kind of an event that we have had yesterday and today is important for a variety of reasons, but it is an example of a lesson many of us have been learning over and over, that is, of the importance of engaging the public in the deliberations of the scientific community and the utility of it.

It is not just about who pays—that old adage, he who pays eats—but it is about taking advantage of the experience and expertise that people who live with these issues bring to the table. I believe, and actually editorialized in *Science* magazine and had no effect about this issue, that it is important to provide the public with an opportunity to help shape the research agenda in a positive sense of shaping.

An obvious question is how do you do that? There are a variety of mechanisms that people have used. NIH has long had a distinguished history of bringing the public into research agenda-shaping kinds of exercises, but this is an example of one, I hope. By having the major funders in the room and the people who get to articulate the research agenda, I think we have a special opportunity, and I am very pleased that we have been able to have as much interaction as we have had.

The topic of public–private partnerships tracks with that, and is a part of it. So with that little bit of context, Sallie Bernard, you're on.

MS. SALLIE BERNARD
SafeMinds

I made the observation last night when we were leaving for dinner that the last time I was down at the IOM building, I was actually outside protesting. Obviously this meeting is very different, and I think we can attribute that to this whole process being a true example of a public–private partnership, from the planning of the meeting to the meeting itself, and hopefully after the meeting with your leadership, and Dr. Leshner in guiding the meeting, and the tone set by Dr. Raub, we really have a public–private partnership right before us in this meeting. I think this is an example of what happens when the advocacy community, scientists, and government come together. That is how we are going to get our environmental agenda in effect, by continuing the path that we have started down.

When you look at public–private partnerships, I see them happening in two ways, at least as it comes to promoting an environmental agenda in autism. One is bringing together people with diverse experiences, expertise, and perspectives in critical thinking, and you need the public and the private sides in order to do that. So it is really a human capital idea.

The second component is a funding component, but you also need to bring people together in forums like this, but it carries across to other activities. For example, if there is a large study that is being designed, bring in the advocate community to help you design and implement and interpret the results of that project.

Things like being involved in the autism strategic plan which we just spoke about earlier today, bringing advocates into that role, making sure that they have a prominent position on the Interagency Autism Coordinating Committee. This is an example of public–private partnership. I think that the organization I represent here, which is SafeMinds, we have certainly tried to historically have a role in those types of activities, and bringing the environmental side of the gene–environment equation to the forefront of autism research.

The second area in public–private is what projects are you actually funding in the research arena? Where SafeMinds has played a role in that is to take a new high-risk hypothesis, and certainly one that is not popular, and people would prefer not to study it, which is the role of mercury and vaccines and thimerosal in autism, and broader than that, the environmental side, bringing it to the forefront in autism research.

To get that off the ground, since the year 2000 we have funded studies on thimerosal and mercury in mouse models, primates, looking at what goes in the brain and other tissues, behavior in primates, cell culture studies, a baby hair study that we did. These are all very small-scale studies, but they help to set the stage, provide a platform from which larger things can come. Now we are at the stage where we would like to see these types of ideas go to the next level, and for that we need the public side of the public–private partnership.

If you are looking at the data and the data say that there is a gene–environment interaction that is involved in autism, and you look at what has happened with autism research, all of the money except for maybe a little tiny proportion has gone to the gene side. So we need to rebalance that. In fact, we need an overcorrection, because we need more money on the environmental side now to compensate for the fact that for the last 15 years, as Laura Bono said at the beginning yesterday, nothing has been done on the environmental side. So we would like to see a lot more of that happen right away.

The last thing I would like to point out is, when we walk out of this room we are going to get a lot of push-back in implementing the ideas that we are all going to go around and have today. We need a mechanism to help us move forward, to make sure that the ideas that we come up with today get implemented. I hope that turns into another public–private partnership and we put that mechanism in place.

DR. HENRY FALK

Coordinating Center for Environmental Health and Injury Prevention, Centers for Disease Control and Prevention

Thank you very much. This has been a really good meeting, in the sense of having an opportunity to listen and exchange information. I want to thank Bill Raub and the organizers for inviting me. Much of my comments will be within the context of public–private partnership, because I don't think there is any potential for success without that.

At CDC, in the last several years the terrorism part of the agency's budget has gone way up, and everybody worries that the bulk of our other public health activities are under a lot of stress. We have taken to using the terms that there are "urgent threats" that we have to deal with and but also "urgent realities." These are Dr. Gerberding's words.

I think "urgent realities" resonates with me in this setting. Autism certainly fits. So, what can we do in this situation that would be helpful and positive, maintaining an open mind and attitude and what can CDC scientists do to help, particularly in the environmental area? As you heard from Larry Needham and others, there are potential ways in which we can assist.

One of the things that I heard over the last day and a half is there are real limits on the amount of work that has been done related to the environmental aspects of autism. I think environment has been underrepresented; perhaps also aspects of epidemiology are underrepresented. I heard a number of issues that are not yet fully addressed on the epidemiology, but certainly more on the environment.

When we look at different ways to address the environment, to me there is always the issue of how do you find the right clues. We don't at the moment have the hook into what are the key environmental issues.

I always see this as two very distinct ways of going about hunting for clues. One is, hit it with everything you've got. When we talk about things like the National Children's Study, we use multiple endpoints and multiple risk factors, we get everything we can into that study to be as efficient as possible, and as broad as possible, and we hope that as David Schwartz says, you look at as much as you can, and you hope that you catch all the right clues. So that is one approach. Larry Needham in the lab has been involved in working with the National Children's Study. We would like to participate in ways that we can contribute to that approach.

I would also like to emphasize the second approach. Early in my career I spent a lot of time with Dr. Robert Miller, who headed the childhood cancer epidemiology program at NIH. He would do things like catch me at a meeting in Japan and say, just look at all those people. The Americans have stripes in their ties going this way and the Japanese have stripes going in the other direction. Somehow you have to really look carefully at the problem to find the clues.

This second aspect requires looking intensively at the cases themselves and, for example, using case-control studies. You have to look for anything uneven in the data, and for geographic discrepancies. You need to look at more severe and at less severe cases and see what can you learn from them. You look at isolated cases that occur someplace else and see what you can learn from those. If there are groups in the population that are not being vaccinated, study them. Whatever the issue may be, search for unusual features and see whether you can learn something interesting that then provides you a broader clue.

One is a macro approach and the other is a micro approach. We need to do both.

I also think that there are many issues here that people have expressed or feel strongly about that are potentially amenable to analysis, and we should try to do as many of those as possible. Whether that relates to chelation or other ideas, if there is some way, for example, of actually analyzing issues like excretion, equilibrium, and metal distribution in children with autism, we should look into how we can best do that.

How to go forward. I think there are lots of good ideas that have been put on the table in the course of the last day and a half. My sense is, we should try to capture the moment so that we not only have these good ideas, but that we figure out a good mechanism, as Sallie Bernard said, to follow up on these good ideas, for example, through a smaller workgroup of people here, to say how many of these good ideas can we put into practice and how can we do that?

But I think certainly on the environmental focus and on some of the epidemiologic issues, there is potential for good collaboration between CDC and NIH within the government, as well as with private groups and others on the outside.

Somebody asked Phil Landrigan yesterday how the lead program made so much progress. There were some key moments in there, one of which was how to utilize data to get government agencies to move forward, for example, how to get HUD (Department of Housing and Urban Development) engaged to remove lead from housing. So, we should think about how to seize a moment like this and go forward in terms of follow-up to this meeting. We hope to contribute to doing that.

DR. GARY GOLDSTEIN

Kennedy Krieger Institute

Thank you. I am a founding and volunteer board member of Autism Speaks, which is now about 2 years old. I thought I would take my time to tell you a little bit about what we are doing.

The object here was to raise awareness of autism in the general community, not only so we pick up cases early, but so the general public would support the research and advocacy efforts of this community. I think it is off to a good start. In the past 2 years, three other organizations funding research in autism, AGRE, NAAR, and CAM, those people in this community would know those, have all come under—have merged in, which is pretty unusual in my experience, of nonprofits merging, and have come in under what is now Autism Speaks.

This foundation has the objective to do as well in fundraising as the March of Dimes or Muscular Dystrophy Association or whatever, and have enough money to augment and influence and create progress in understanding and treating autism, preventing autism.

I will give you some examples of the infrastructure things that have historically and more recently have been funded or are going to be funded. They are things like the Baby Sibs Consortium, which is a group of about six to eight different centers now looking at the subsequent children born to a family that has one child with autism. There are lots of interesting observations that can be made. But the idea of bringing this group together and sharing data.

Second is something called the Autism Tissue Project. This is one to encourage—when a child with autism dies, to encourage a donation of brain tissue that can be banked and used for histologic study. We greatly lack careful and detailed studies of the anatomy of autism.

A third is the AGRE Project, which is the multiplex family project of collecting genetic material and making it available to investigators widely, with the idea that again, the data are shared.

Another one is the IAN Project that you heard about before (http://www.ianproject.org). I would encourage you to look at this. It is a community site and it is a research site of online research. One big goal of that is to match and provide potential subjects for studies that someone is appropriate for and would like to participate in.

The Autism Treatment Network is under discussion right now. This is benchmarked to the successes of the Cystic Fibrosis Network. We are at a much earlier stage. We would also like to have the ability that every child in this country who has autism can reach a medical clinic by car. We in Baltimore at the Kennedy Krieger Institute are continually getting referrals, people who are flying in from different cities for what seems to be very simple evaluations. They should be able to get them in their own community, but they can't because there is no one there to provide them, or the waiting list is so enormous.

So one of our goals is to have treatment centers, medical centers, in children's hospitals hopefully that can evaluate children within a car drive, an hour car drive, and that these people share standards of evaluation, standards of care and their observations are collected and centralized in a data management system.

These are some of the infrastructural things we are funding that I think will be available for research, rather than specific projects. We do have projects. We have pilot grants, fellowship grants, and most recently augmentation grants. CDC is funding a project, and it looks like something more can be done with an augmentation grant, that we would be open to that, or the NIH has a funded NIH grant, or a grant that is a near miss. We are finding now the grants that have outstanding scores, priority scores of 140, not getting funded. I think autism is not in a separate pile. We would be open to helping that investigator keeping his or her investigation going via a new R01 or a continuation grant that just didn't get funded until they have time to go through the 9-month, year process to get that grant reconsidered.

So these are some of the things we are doing. Then we are looking at gaps. I have tried to do a portfolio analysis. I was able to collect the amount of funding from each of the institutes directed to autism projects. These were projects that had autism in the title. I could see where the money was going roughly.

The gaps that we talked about. One is, if there is a role for immunologic mechanisms in either the treatments or the prevention or the cause of autism. I only could find one grant. I found 0.6 of the NIH budget going to the immunology of autism. NIH doesn't plan what it spends, this is just in response to what is submitted. That is what was submitted and that is what was funded.

Lots of money going to gene searches, defining autism in children. Very little to drug studies, but then there are no targets, so maybe that is why that is the case, but very little, and very little going to the whole toxicant things that we have been talking about, very, very little.

So as we looked at the gaps, these are gaps we see. Are we going to help fill them? Are we going to partner with different agencies to help fill those gaps? Are they appropriate gaps to work on right now? That is what Autism Speaks is about.

DR. TOM INSEL
National Institute of Mental Health

My interest in being here was generated through the evaluation we did for the IACC autism research matrix.

Just to back up for a minute, there are five NIH Institutes that contribute to autism research. For the most part, all served on the IACC that was enacted under the 2002 Children's Health Act and which included 12 federal agencies as well as public members.

At Congress's request, in 2003 we developed an autism research matrix which was a set of short-term and long-term goals, both high and low risk, over a 10-year period. In 2006 we reviewed the matrix, including fairly broad input in our analysis. Most recently, the matrix was out for public comment in a phase that ended January 16.

In putting those comments together, our expert reviewers, the people on the IACC itself, and the public all gave us very much the same message: You have done some things very well, such as building capacity in certain areas, particularly genetics and some tissue banks, and you have built up the field in a number of ways. However, you have not paid enough attention to environmental factors.

So going forward, I think it is fair to say that is going to be an area of increased interest at NIH. I'm sure Dr. David Schwartz, who is the expert in this area, will fill in more of the details.

In terms of the public–private partnership, I think it is important to realize that autism is different from many other areas of biomedical research in that we do have a very significant private investment here. Autism Speaks is aiming to have about $60 million a year to invest in research. The Simons Foundation has already put in about $30 million and just announced a new RFA last week for another $15 million investment this year. This is in addition to the $108 million NIH is spending. It is really more than just an exchange of ideas; there is a real opportunity here to put in money that will synergize efforts and perhaps serve as a catalyst for certain things we want to accomplish.

The question is—and this goes back to what Sallie Bernard started with—how do you divide this up? At the intellectual level, there is an opportunity here for a much better partnership. You heard a lot about this from Laura Bono at the very beginning of the meeting, such as the expertise that families can bring to this discussion and to this partnership. Then the question becomes, how do you facilitate that, how do we make sure that we are involving families as partners as well as these other private groups?

A second question arising from Sallie Bernard's division of this into intellectual and infrastructural or resource-driven money, is how we can make sure that we are working together and not at cross-purposes. Is there a way that in many of these areas we can divide and make major progress, and say Autism Speaks will take on this challenge, CDC takes on a different challenge, NIMH addresses another part, and by following the expertise, get the best science done? I bring this up because there is a lot of energy in this room and a lot of enthusiasm for opportunities on environmental factors, and this has been great to hear. That is what I wanted to get out of this; where are the opportunities, where do we want to move, how are we going to make this happen?

We haven't talked much about the challenge of peer review and what happens when many of these ideas, some of which we talked about here, go to peer review. Innovative and cutting-edge science, particularly ideas for which there are limited pilot data, often face significant challenges through the peer-review process.

So we have this tension between the sense of urgency that many people in this room feel and the culture of science, which ensures the rigor we expect, but can also make it very difficult for us to move quickly and sometimes makes it more difficult to move in new ways.

So here is my suggestion. As we think about public–private partnerships, we think not only about opportunities and who will do what, but also in terms of the kind of science we want to undertake. As Dr. Gary Goldstein was mentioning, there may be an opportunity here for particular areas, which might struggle in peer review, to be picked up, at least for the pilot phase, by Autism Speaks or the Simons Foundation or by some other group before they maybe get taken to a larger scale and get funded by NIH.

I don't know any other way to do some of the most innovative projects, especially at the earliest phases. So I think that is the public health challenge, balancing urgency and rigor. We just have to get our heads together to figure out how to make that happen.

MS. LYN REDWOOD

National Autism Association

First I want to thank Dr. Raub with the deepest gratitude for listening to the parents and taking our concerns to heart and making this important meeting a reality. I want to thank Dr. Schwartz and Dr. Insel and the staff and the panel participants for taking away time to commit themselves to this important research. I hope this will be the first of many meetings in an effort to get to the bottom of what is making our kids sick and what we can do to help them now.

This is a mock newspaper article. If tomorrow we woke up and 1 in every 150 children were reported missing by their families, what would happen? It would be a national crisis. Communities and federal agencies around the country would band together to form search parties in an all-out effort to find the lost children. I am asking today that our federal agencies move into crisis mode and rethink the way they conduct research by reaching out to the persons who are the most knowledgeable about this disorder—parents.

Parents are confronted on a daily basis with the needs of their child. They are the ones who are most knowledgeable about their illness. They also have the highest level of commitment to finding help, because anything less is just not acceptable. Parents are able to add perspective, passion, and urgency to this discussion, assuring that the human dimension of the disease is incorporated into scientific considerations, program policy, the investment strategy, and research focus.

With PubMed available online, parents can now access research that in the past was only available at medical school libraries. This has tremendously increased their knowledge base of the disease and has leveled the playing field between parents, physicians, and researchers.

An article appearing in this month's issue of the *Archives of Pediatrics and Adolescent Medicine* documents the important role of parents in autism research. Parents have organized research funding, they have constructed clinical research networks, they have popularized empirical-based treatments, they have suggested new avenues for research, and they have anticipated shifts in the understanding of autism. The article concluded that the existence of partnerships with parents is a critical component of future research and treatment programs.

I would like to outline very briefly four different mechanisms that are available now to be able to partner. These include autism advisory boards, integration panels, the establishment of shared research inventtories, and community-based and -driven research initiatives.

In the Combating Autism Act, Congressman Barton called for the creation of an autism advisory board to provide public feedback and interaction with the NIH Interagency Autism Coordinating Committee. He went on to state that public participation is necessary to emphasize the human side of autism research, and to ensure that federal resources are used widely.

An idea to embrace consumer participation was also recommended in 1993 by the Institute of Medicine in a report solicited by the Department of Defense in the development of a congressionally mandated research program for breast cancer. In this program, consumer participation occurs at both levels of peer review, scientific and programmatic. Consumers read proposals, they present their opinions, and they have full voting privileges on the committee. A recent published review of the program found consumer participation a very positive and successful aspect of the program, and suggested that it be used as a model for those who desire to work in partnerships on critical health issues.

Parents are not waiting for the double-blind, placebo-control studies to determine if a simple intervention such as changing a child's diet, administering supplements in an effort to support the child's nutritional status, or reduced oxidative stress is beneficial. Parents are leading the science beyond educational or behavioral interventions because they recognize that their children have medical problems which present with behavioral manifestations. Addressing these medical issues often results in improved overall function for the child.

The Autism Research Institute reports that over a thousand parents have completed questionnaires regarding their child's recovery from autism. In addition, several large clinical practices that utilize a biomedical approach to treating autism have offered to open up their practices for data mining to identify historical, physical, and clinical information which could provide clues to autism's etiology, heterogeneity, and effective treatments.

In summary, we are faced with an urgent public health crisis that demands immediate attention and action from our federal agencies and the allocation of resources necessary to respond rapidly and effectively. This change in paradigm dictates a shift in the focus of autism research away from an exclusively genetic model to one that investigates the role of environmental factors, an abandonment of the traditional bed-to-bench approach to research that is predominantly investigator driven, to one that embraces a sense of urgency and direction infused by parents.

Parents as stakeholders need to have their status elevated and voices heard. We need mechanisms where input and ideas are actively solicited and formally embraced by our federal agencies, one where stakeholders are able to guide research in a direction which offers the most promise in an effort to find meaningful ways to help improve the lives of those suffering with autism now and to prevent its occurrence in the future. We need to devise ways to partner, share ideas and resources, and work together in the areas of policy, science, and research.

Thank you.

Dr. Leshner: The last speaker on the panel is David Schwartz.

DR. DAVID SCHWARTZ

National Institute of Environmental Health Sciences

My view is pretty simple. Autism is a complex public health problem that we are not going to solve from any one perspective. Therefore, it seems to me that partnerships, including scientific partnerships, public–private partnerships, partnerships with families, and partnerships with patients are essential to being able to understand this complex problem.

In fact, I think over the past couple of days, what is clear to me is that without understanding the genetics component of autism, we are not going to be able to make headway in the environmental component, and without understanding the environmental component of autism, it is going to be harder to make sense of the underlying genetics.

That is probably why 15 of the 23 pairs of chromosomes have loci that have been linked to autism. There must be lots of different causes of autism, lots of different phenotypes, and I think the environment and other factors can help us understand the causes of these different types of this disease.

What are the fundamental bases or ground rules of a partnership? The ground rules of a partnership are that everyone wins from the partnership, that we all make each other better by being part of this partnership. It seems to me that autism and this problem that we are dealing with is just the kind of mission-oriented problem that would benefit from input from several different vantage points and individual parties or groups working together to solve this complex problem.

Plus, I think it would be a lot of fun. Over the past day and a half, I have enjoyed time thinking about this problem with everyone around the table. It has been enjoyable to me to consider the ideas that people brought forward in the context of scientific questions.

Now, I don't want to reiterate the scientific gaps, because I think we are going to get to that later. What I want to do is bring up three infrastructure issues that I think are important to consider in a partnership.

One is, I think this idea of a biobank that cuts across studies might be a helpful way of creating transparency, creating access to data, and creating access to samples both today as well as 5 years from now that could be helpful in solving the problem.

A second infrastructure need that seems to me to be apparent is, when I think back to the early 1980s when I was taking care of patients with AIDS, we were all running around giving different types of therapies in uncontrolled ways. It seems like therapy for autism is at risk of a similar type of poorly coordinated, poorly studied problem. A clinical trials network would ultimately benefit children with autism, if patients, families, investigators, and clinicians embraced this approach. I believe we could make the same progress that we've made with AIDS if we develop and support a clinical trials network for autism. AIDS patients were leaders in the AIDS clinical trials network and I think the same could be done with families with autistic children.

The third infrastructure need that I think is important to consider is training in this area. While I think that experienced, established individuals could contribute to this area, I think we need to think about the next generation. So I would want to think more about the possibility of developing environmental sciences programs that could inform the behavioral sciences, cross-training individuals so that they could develop new knowledge in this important area of research.

DISCUSSION

Dr. Leshner: Let me open the discussion to people at the table and see what your reactions and your thoughts are. Tom Insel answered my question that I had for Dr. Goldstein, which was going to be about how much money are we talking about from a variety of organizations. It would be really interesting and perhaps as a next step coming out of this workshop to have serious sitdowns among the various funders about how to do a better job coordinating and parsing out who does what, because we are talking serious money. They are not trivial amounts of money coming out of the private sector, whereas in some other domains there really is trivial money coming out. So there is something worth coordinating.

Dr. Insel: There is a conspicuous absence though in this discussion, which is pharma. In almost any other area of medicine, if we were talking about such an effort we would be talking about pharma, which represents double the investment of the NIH every year in R&D (research and development). So one might ask whether that is also something that could also be brought into the discussion.

Dr. Leshner: Can I just ask out of ignorance, do pharmaceutical companies do research on autism?

Dr. Insel: No. According to them, there are no targets. They don't think the market is big enough.

Participant: And it is children.

Dr. Leshner: Comments?

Dr. Spence: I just want to say, pharmaceutical companies are running a few trials, a little. I don't think they spend the kind of money you guys are talking about, but I don't want it to go on record as saying they are not doing anything. Existing psychiatric drugs are being tested in autistic individuals.

Dr. Leshner: Since people aren't jumping to say something about this public–private thing, I would like to challenge the group, that what we have seen yesterday and today is actually a very good example of a public–private partnership, and the beginning of a process of conversation and a model for perhaps how to have these kinds of interactions. I really think we need to give Bill Raub a great deal of credit for having pushed this as hard as he did, which was hard, but well as always.

But I would say that since we have the making of a public–private partnership, if we don't have one going forward, shame on us. I don't actually know exactly what I am talking about in terms of how to make sure that it has momentum and continues, but at a minimum we have the base for it, and maybe a next step to assure that it partly continues is to do two things.

One is to try to get the funders together, because that is always a good way to get things moving. The second is to have some discussions among the organized groups of things like what the scientific criteria would be for acceptable kinds of studies. I think that is something where there is some either disagreement or confusion among the various groups, about what actually is a fundable study and what is a doable study and what will have the credibility or not the credibility.

Again, that needs to be a two-way discussion that goes on between the people who want an array of new kinds of research done. It may have been Tom Insel who raised the issue of peer review. I can tell you as a journal publisher that 96 percent of the studies that come to *Science* magazine get turned down, but an awful lot of this kind of research would have a terrible time getting through normal peer review processes.

I have to ask, where do you move the line, where do you risk moving the line? I can promise you, from my own experience in the drug addiction business, if you lower the bar, the bar will fall. So we will have to be very, very careful, because it is a very fragile bar, much more fragile than people may think. So I just offer that as unsolicited advice.

Having said that, let's open up for the around-the-table gap business, what you didn't say at lunch. Alison will start.

Ms. Singer: I'm going to do two in 30 seconds each. I want to talk a little bit more about the time trends data that Mark Blaxill brought up before, and say that this continues to be a dark cloud that distracts us from the real issues that we have to solve.

To Tom Insel's point earlier about what will we be sorry about when we look back 5 years, we need to do the retrospective studies, and we also need to start to gather the data now in the ADDM studies, looking at the subtypes, what percentage of the kids are Asperger's, what percentage are autism, and what percentage of the kids are PDD-NOS, so that we can use that to inform the science and also use that for planning purposes so that we can plan for treatment for all of the kids. The treatment protocols across the life span for a person with lower functioning life-span autism will be dramatically different from the treatment protocols for someone with Asperger's.

I also want to say that we need to focus on studies that yield actionable information for parents. All across America there are thousands of minitrials going on every day in our homes, where we are experimenting on our kids based on anecdote, and we deserve better. We deserve better than anecdote, we deserve evidence.

I would urge when we are looking at the list of 80,000 chemicals, that we focus on those that are still in our environment, that are still being used, where the data that we glean from the studies will result in actionable information.

In my last 10 seconds I will say by way of an example, I don't think that many parents or pregnant women are still prescribed thalidomide, but they are still prescribed terbutaline. So I would say, let's try to prioritize in a way that helps parents to have information that can help to prevent more children from being diagnosed with autism.

Dr. Colamarino: I don't think this will come as any surprise. I would say biomarkers, biomarkers of exposure, biomarkers that are predictive, biomarkers of outcome, especially treatment outcome, and biomarkers of change across treatment lifetime. I would qualify that as saying physiologic biomarkers, measurable things.

Dr. Needham: And biomarkers, but in addition to biomarkers, and this has been touched upon also, is a need for biobanks. But importantly in that is the correct procurement and storage of the specimens. There would be nothing worse than taking samples, and 5 years from now the samples are neither taken nor stored correctly. So we need to talk to each other about how to take and store those samples.

Also, we need to look at other stressors, certainly environmental chemicals come to mind, but we also need to look at such stressors as infectious agents and also the effects of the psychosocial. There are many stressors that may play a role in autism.

Ms. Bernard: I agree with everything that has been said, so I won't spend time on those. I would also like to add the recovery study, which we could do with videotapes of when they were autistic to verify that they really were, and then when they are not recovered and have a blinded—we could do a study like that and we could do it right away.

I think we also should do a study of cases of unusual onset, because they give us clues. We talked about first trimester or whatever, but there are also cases in the literature more than those of people getting autism when they are teenagers, when they are in their late childhood, from viruses or other infectious agents and from mercury exposure. We should look at those, and maybe it gives us information on windows of vulnerability.

We should also look at adults, not just with biomarkers as you said, Sophia, but also phenotypes and how that changes, and what do teens and adults get that the children don't have.

I would also have us consider when we do studies of low-dose and paradoxical effects, because that seems to be something that affects studies when we are looking at dose response, to consider those.

Then something that is related to research, push maybe a little heavier on the precautionary principle, and look at what are the top substances out there that could be causing harm and raise the alarm that we really need to do something about all these things in our world that have a high possibility of making our kids sick.

Dr. Goldstein: I'd just like to say that we are learning from the IAN Project that more than 80, probably 85, percent of parents have never before participated in a research trial. If you contrast that with what goes on in the successes of childhood cancer, where over 95 percent of the children are in research trials, we are almost the reverse. And these are people interested. So I think there is an enormous difference between what is going on in terms of testing anything, whether it is diagnostic tools or whatever.

Then I would add that I think this would be a good time to begin to explore the role of the immune system as it relates to autism.

Ms. Redwood: I think it is really important to go into the database of the children, as Sallie Bernard said, that have recovered, but also the ones that are currently undergoing treatment. I think there is a wealth of information there with regard to biomarkers to be able to better identify the heterogeneity of this disease.

I also want to put a plug in for—and I missed this in one of my slides—having public and private agencies to develop some type of method of coordination. As we have heard, the private sources are matching the public sources now, and may very well exceed the public's resources in terms of autism research. So we need to coordinate those efforts so there is not duplication in the areas that Gary has mentioned that have been underfunded in the past. Historically, environmental research comes to the forefront.

So again, I am echoing things that have been said previously, but I think they are very important.

Dr. Schwartz: I feel like I have made too many suggestions, but I'll make one another, which is the data access, coordination, and sharing policy among autism investigators, as well as a vehicle to make that happen.

Dr. Schendel: Since one of the driving forces behind this meeting is our hypothesis that environmental factors contribute to the changing prevalence of autism, one of the things that we lack is an understanding of the features of individuals who were born in the 1980s.

So we might want to consider a descriptive study of individuals in their late teens and early 20s, and if it was population based you might even get a better handle on prevalent characteristics.

Dr. Newschaffer: I'm pretty confident that most of my personal high-priority areas have been addressed or will be addressed. I want to go to a couple of areas.

First, in relation to the topic I was asked to speak upon, which is the international studies, especially in developing countries, I think we need to work on culturally robust, easy-to-implement screeners. I think these may have to rely more heavily on child observation and language and interview, because the combination of language and culture make it very difficult to do screeners that are effective in developing countries. That is one area.

Second, I want to drill down a little bit in the area of retrospective studies for time trend. Maybe we need to involve folks from other disciplines, like medical sociology. It is very hard to figure out how to evaluate empirically diagnostic tendency retrospectively. You can't look at what is written down, you can't look at records, you can't ask people what they think they were doing 20 years ago.

When you examine the older kids, which is a nice idea, the phenotype has changed, and if you assess them now, you are assessing them through the current lens, not the lens of what evaluation was 20 years ago.

So perhaps we need different disciplines to think about this, to help us come up with valid ways to do this kind of study.

Dr. Hertz-Picciotto: We were talking about that at dinner last night. It is definitely an issue that needs to be addressed, including the possibility of people having been institutionalized and therefore having undergone heavy medications along the way. So it is a difficult question, but we do need to figure out if this is a feasible thing, to go back and see the autistic adults, if they are out there.

The other issue that hasn't been brought up and hasn't really been talked about, I think there might be something that can be learned from looking at the other things that have been increasing in this same period of time, if there is a true increase, and there may well be.

I am thinking of asthma as one, which we heard a bit about, and also the obesity epidemic. I think those may point us in the directions of immunologic and metabolic things, that perhaps there is some commonality here. Obviously very big differences among those things, but something that we may be able to learn from that.

Ms. Bono: I think I said some of mine earlier, so I am just going to run down them very quickly, and I'll stay within a minute.

I agree with Alison Singer and several others that time trends are very important, what changed in the environment and when. The phenotype is very important, biomarkers that Sophia Colamarino mentioned, the clinical data mining Lyn Redwood mentioned, recovery studies that Sallie Bernard has talked about.

We discussed this morning, and I am going to bring up again, we need a vaccinated and unvaccinated study. It is very important to the community to be able to move forward. We have to look at that, we have to move beyond that.

We need a mercury toxokinetic study, especially because so many kids are getting better. They do have heavy metal problems, and we need a chelator that will cross the blood–brain barrier. Currently, and this is maybe where pharma can get involved, we have chelators that were created in the 1940s. I know we can do better. It is 70 years later. I'm sure we can do better.

We need methylation detox types of studies. We need anti-inflammatory immune studies, which I think Gary Goldstein brought up. We need gastrointestinal studies, including the enzyme systems and bacterial gut viruses.

I love that Dr. Insel brought up peer review. I think that is a big area we need to discuss, especially about scoring proposals, based on the gaps that we are seeing here; put the priority on the studies that we need.

And bottom line, which I know we all are thinking about now, that is helping the most kids as quickly as possible. If we keep that in mind, then I think we are going in the right direction.

Dr. James: I'd like to address the issue of biomarkers and the biorepository, which we have all alluded to. There are several epidemiologic studies that are now going on collecting biological samples, but I think there are important practical considerations in the details that are absolutely essential for sample collection.

I think we need to establish standardized protocols for sample collection. For example, for our glutathione redox ratio, I can't use your samples. It has to be fasting, it has to be at the same time of day, dumb little details that are going to make all the difference in being able to have meaningful data that we can compare between studies.

Another big area that is a problem for me is age-matched control samples. If we want to find out what is unique about autism with these biomarkers, we have got to have something to compare it to, and they are very hard to come by. I don't know how to approach that, whether we can have a control repository at the same time.

The other practical problem with the repository, especially metabolic biomarkers, is that they degrade over time. They have to be stored properly in the same way. So I would just make a plea for standardized protocol for collection. It will take a small working group to establish those criteria.

Dr. Noble: Three things that haven't been said. On the immune system, I think Gary Goldstein is right. I think one can make the prediction that these individuals are in TH-2 imbalance. They will have low glutathione levels. It is easily studied in T-cell populations. They will have memory cell defects. There are a lot of data from the AIDS literature on the effects of glutathione decreases on those areas. It might be useful surrogates, if not functional.

Isaac Pessah said we can't get access to the nervous system. I have been thinking about that. What we can get access to is neural crest stem cells. They are very easily grown from skin biopsies. That has been very well worked out. That may be something for us to look at.

Lastly, I made the statement earlier that this is unlike any other neurological disorder that I have been interested in in terms of these recovery stories. That was a misstatement. This sounds like a biochemical disorder. That is where we get those kinds of turnarounds. When we get the right drugs into an individual who has a biochemical disorder, we have these effects. So I am starting to now go in that direction in my thinking for awhile.

Dr. Goldstein: In terms of neural tissue, we do nasal biopsies in Rett syndrome, and can get growing neurons from nasal biopsies.

Dr. Noble: So you and I should talk about what would be the best to look at.

Dr. Beaudet: I guess I am the genetic representative. There seems to be a certain sense that maybe genetics is somehow competing for resources with the environment considerations. I would just make a couple of points about the genetic situation.

There are dramatic advances going on at this moment. They are not at all the outcome of the money that was invested. They do with discovery of *de novo* defects, and all this big investment has been in inherited kinds of abnormalities.

I think this group that have *de novo* defects are by and large mentally retarded and dysmorphic, and they can be identified to a great extent and moved apart from some of the other studies, because I think they will make any other study more complicated and contaminated if it is not done.

Another thing which I haven't pushed particularly is epigenetic sampling, which is very difficult because of the tissue specificity. I have a sense there is probably a group of patients where the etiology is mostly genetic, and another group where it is mostly environmental with some genetic susceptibility component. I think it would be helpful to separate these.

Then an area where I don't really have any expertise, I still feel like we need to know whether the incidence has changed between 1970 and 2000. Even if it is stable now, we need to know if it changed.

Mr. Blaxill: I want to talk a little bit about the burden of proof on time trends. I would make the suggestion that given the increases that we have seen, the notion that the reported increases are an artifact is a hypothesis, and it is a testable hypothesis.

I'll just take California as an example, because there is a pretty good surveillance system there, better than other parts of the country. A child born in California in the early 1980s had less than a 5 in 10,000 chance of becoming autistic. By the late 1990s, that rate was closer to 40 for 10,000, so that is roughly a 10-fold increase in about 15 years.

The notion of that increase being artifactual has been tested in a lot of natural experiments. There is a hypothesis of diagnostic substitution that has been tested and falsified. There is the hypothesis of diagnostic expansion, that somehow we are changing the quality of the diagnoses.

The interesting thing about California is that the registry is for a full syndrome, it doesn't include the broader spectrum, so that theory of expansion doesn't hold. The M.I.N.D. Institute has done a quality control check across decades. There are problems with those kinds of studies, but they didn't uncover any different diagnostic quality in birth cohorts from the 1980s or the 1990s. And the surveillance system has been in place. It has changed, any administrative system changes, but it has been in place since the 1970s, unlike some of the educational records in the 1990s. So the notion of diagnostic expansion is not supported.

The only remaining hypothesis, or what I like to call the hidden horde hypothesis, that somehow hundreds if not thousands of children escaped the service systems, and if we looked for 25-year-old Californian young men and women with autism, we would find them in large numbers somewhere. That is an interesting hypothesis. I would suggest people ought to prove it before we start accepting the notion that the increases are artifactual.

So all you can say from that is that there is a lot of evidence that suggests that in California the increases are real. California, when you compare other databases to the rest of the country, they don't look that different, so again that is inductive reasoning, but you could argue that is a pretty useful database for the United States.

Then I would ask the question in terms of studies, I think we should pursue studies to clarify uncertainties, but I would urge us to consider changing the burden of proof. Rather than saying the burden of proof is to demonstrate that all this is real, I would say the burden of proof is to demonstrate that it is artifactual.

If that is the case, we ought to think about changing our official narrative, because the expression of doubt about the increases creates the sense that we have a mystery and a puzzle, and no sense of urgency. The recognition of the reality changes the entire dynamic. I think a lot of us are saying we need to treat autism as an emergency, and that is what all the data points to.

Dr. Susser: I agree that we should really tackle the time trends problem, and that we should trace the course of autism over an individual's life span, not just over the first few years. So I second those two points that have already been made.

The only point that I would like to add is that I would like to extend what Art Beaudet said, but in a different way. I think one thing that we have learned from all these genetic studies is that there is a sizeable proportion of kids with autism that do have *de novo* genetic defects. That is really important, and that is new knowledge.

I think what that should lead us to do is to look for the causes of those *de novo* defects. They are probably going to be environmental. I think we should focus there when we are thinking about environmental studies of autism.

Dr. Landrigan: On the first day of science class, we are generally taught to beware of the study with the *n* of 1, and for good reason. But to summarize from yesterday, I think we heard several instances where the *n* of 1 study, to the extent that it is a hypothesis-generating process, can be extraordinarily valuable, especially in the face of a disease of this complexity.

But also, it would be a way of ensuring that the engagement of the parents and the treating clinicians and the scientific community come together. The problem is that peer review will be deaf on most of those kinds of studies seen one at a time. But if we could create an infrastructure, direct the peer review to the mental level in terms of something that fosters those opportunities, and then make it easier somehow, a lower energy barrier, whereby opportunistic examinations of that sort, testing a variety of the hypotheses we had here, I think it could be seminal in terms of shaping the more deliberate and longer term research protocol.

Dr. Herbert: That is really great, thanks. I want to mention three quick points.

I think there needs to be support at the infrastructure level. Sallie Bernard talked about makeup funding for environment. I think there needs to be makeup funding for parents and integrative practitioners to come together and figure out what it is that they need to convey about the disease phenomenology and the methods of measurement that should be fed into the biomarker development and study design development process.

The second thing is, there has been a lot of talk about having some kind of a biomarker consensus meeting or think tank to come up with standard operating procedures and standardized measurement. I think that should be fast tracked.

Finally, with regard to interesting existing scientists and companies in studying autism, which is not a huge population, there are a variety of conditions which have overlapping biochemistry and pathophysiology with regard to inflammation and oxidative stress, various neurodegenerative diseases across the life span, obesity, diabetes, and work has been done in those domains. People who have been working in those domains, that could be recruited to this effort so that it wouldn't have to start from scratch in autism.

Dr. Pessah: A major gap in our knowledge is the neurodevelopmental toxicity of some major priority pollutants that fall below the radar screen of the EPA. These are non-dioxin-like. They have tremendous potency toward certain signaling systems that may be very relevant to autism and other neurodevelopmental disorders. I think we need to promote applications that address low-level exposures that try to understand specific mechanisms that are relevant to autism.

I think we can also use cell samples from autistic kids, preferably primary cells, to try to validate or further understand how these mechanisms are hypersensitive or insensitive to certain mechanisms.

Dr. Spence: I have nothing new. Everything that everybody else said, and one other thing I stole from the father of one of my patients. He said, we ought to be able to look at the brain and decide if there are toxic elements. He said, can you look at the brain and see if there is mercury in them? I said, not with the technology that I know about. But I think the imaging technology is getting better, so maybe there are ways of development of technology to look at toxic exposures directly in the nervous system or in other systems.

Dr. Swedo: This is the advantage of being at the end of the table; we can go fast. I would like to suggest that medical and genetic workups be done on every child who is suspected of having autism. Typically they are sent off to the waiting list for the developmental clinics without having had an EEG done, or other very basic workup. If the child wasn't able to walk, we wouldn't be sending them off for a psychologist to evaluate them.

We need hypothesis-generating research. We heard about the NM-1 studies, but there are some case series that have been ignored.

I would like to see genetics added to all of the epidemiologic studies that we have heard about this morning. If they aren't already being collected, genetic samples should be, and would certainly get at that question of genetic susceptibility to environmental risk.

Common measures across the patient population to allow comparisons. Common measures of assessment, of recovery, of response, but also as we have heard, of biologic measures.

I would like to propose a plan that we have follow-up evaluations for the IAN database. I know that Paul Law and the folks at Autism Speaks are working on ways that those individuals who can't currently participate in research rapidly could.

I will finish by echoing Gary Goldstein's plea that we move very quickly to a cystic fibrosis or cancer-line treatment network. I know that the Autism Treatment Network and others have that as a goal, so they would need additional support to make that happen.

Dr. Zimmerman: I would like to make two strategy points. First, I think there is a base for both a theoretical approach to this, but also an empirical approach. The theoretical approach is, all the scientists are here, and finding a neurobiological answer to why we have this problem.

Second, I think the empirical approach would be a meta-analysis of all the treatment paradigms that have been tried to date, including biomedical, dietary, nutritional, home-based, or relationship-based treatments, because I think there is a probabilistic and causal relationship there.

Dr. Coetzee: I am like Alan, I don't have a particular stake in this disease. I come from the multiple sclerosis world.

In thinking about what I have observed in the last couple of days, there are two areas. One is, I think the cystic fibrosis model is particularly powerful for another area, and that is the foundation's approach to companies, in terms of contracting with them to do research on a population of 30,000.

There is not a big market for cystic fibrosis drugs, but they have managed to invest $200 million over the last 6 years in company research. So I think that is a powerful model for the private sector to stimulate pharmaceutical companies to be engaged in this process. It is not the usual big players, but it is the small innovator companies that will come out of the young people who get recruited to be researchers in autism research and other areas.

I would also suggest that we think about looking at other diseases that may have an adult onset, but that have complex genetics that have environmental triggers and environmental factors for which we don't yet know causes. I think we don't want to necessarily reinvent wheels that may have already been invented elsewhere.

Dr. Cohen: I come here as a member of the Forum. This has all been very new to me. I would like to thank you all for an extraordinary 2-day education.

From that admittedly uninformed perspective, two things struck me. One was the absence of imaging in the discussion. I was going to say fMRI, except Martha Herbert last night advised me that there were other directions to go.

The second thing I missed was a deeper discussion of viral pathogens. Like many of you who have been struck by a lot of the anecdotal conversations, that struck me as well, some of the conversations about the kids with severe GI distress that responds to antivirals, just around the table, the very late onset with possible viral etiology.

While I wouldn't necessarily say that that is my lead hypothesis, it certainly strikes me as a significant element that has to be considered.

Dr. Fombonne: Two ideas. One is to go back to the idea of setting up a study of discordant twins. I think the rates of twins are going up, and that would be an opportunity to look at not only environmental risk factors, but also maybe product risk factors, what makes the outcome different in that case.

Second, it follows up on the last point. I think we should probably look at environmental exposure which might increase the rate of mutations in germ cell lines of fathers in particular. That would be something which would also be to follow up.

Dr. Insel: There is not a lot left at the end of the row here, but let me give you one from Ian Lipkin which hasn't been mentioned today, although David Schwartz started to get there, using microbiomics approaches, that is, this new way that Ian Lipkin described of looking at all the potential microbes through sequencing, not through culture.

I think intuitively that is a great opportunity in this area for a lot of reasons. It fits in with the immune story, it fits in with potential time trends. There is just a whole range of things that would help you understand what is going on, and it is completely untapped. It is now doable. It wasn't doable 6 months ago. But this is a place one might go.

Two others. Mentioned yesterday, not today so much, Sue Swedo started by saying there are plenty of genetic diseases, Mendelian diseases, that have autism as part of the story. If you had anything else that increased the risk as much as Fragile X, you would want to know why. It seems to me that is also an opportunity.

The last thing. Several people have mentioned international efforts. I do think, besides finding these natural experiments and special cohorts, there are a lot of things going on in other countries that we need to be more aware of.

I was recently at a national autism meeting in The Netherlands. Someone came up to me and said, I have been collecting CSF on several hundred people who now have autism. Would that be of any interest to anybody? So there are opportunities out there that we need to be looking beyond our own borders to try to exploit.

Dr. Leshner: That was pretty spectacular.

I'm not quite sure what to do now other than to open it up to the broader floor with the same rule. You get one minute, no more, from the audience. Line up at the microphones. You have got to yell, we are collecting all of this. Name, where you are from, and no speech. A one-minute opportunity.

Session IX
Discussion with Meeting Participants and Audience

Participant: I am from George Washington University. One of the things that has been maybe just hinted at here but hasn't really been addressed is the possibility—and also, this would involve parents and clinicians as well, is the use of response to medications, to phenotype individuals for further study.

As the parent of a child with Asperger's syndrome, he has been thrown bucketloads of drugs, most of which have absolutely no positive effects, lots of negative effects. But if we can tap that database and sort out those kids that may respond well to antidepressants or antipsychotics, those are probably good biological phenotypes that can be useful in genetic or genomic or other studies as well as metabolomic studies.

Participant: Dr. Nancy O'Hara. Two points I would like to follow up on from the panel. One, the University of Maryland has just partnered with Autism Research Institute to start to collect and bank tissue samples, not just brain samples, from children and adults with autism. I think that is very important when you look at environmental factors.

Second, Dr. Sidney Baker, who I think is one of the most brilliant clinical minds in this research, is launching a beta trial of metagenesis, which is a collaborative technological tool to collect data from parents. Ten thousand parents will be receiving letters next Friday.

I offer that to the board to look at that as a means, not of scientific rigor, but as a means of data collection, to start to look at what biomarkers, lab data, treatment protocols might be out there. If you don't receive a letter as a parent or members of the board, you may go to http://autism.com/ to look at that. I think that is something we can use to start collecting some of these data.

Participant: Richard Deth from Northeastern. There is a meeting at the same time as we are meeting here of the think tank from DAN and the Autism Research Institute. I want to bring greetings from them, because there has been some back and forth and there will continue to be back and forth. They represent an organization of resources for cooperation in these kinds of relationships, especially parent based as opposed to fundraising. Different organizations have different foci and so forth.

That group is really anxious—you will probably receive something official from them, offering some cooperation in whatever way possible from that group.

The other thing is, Dr. Insel, you pointed out the importance of private funding because the R01 process is fraught with problems. I, for example, had an R21 where the primary reviewer just cut and pasted the thimerosal statement from the FDA Web site instead of reviewing my grant. So this is really important for the environmental factors that other people step up and never get through the NIH peer review system.

Participant: Mary McKenna, University of Maryland. We are talking here about environmental and inflammatory factors and how they impact on metabolism in the developing brain.

I just wanted to point out, related to what he just said, that brain metabolism studies have difficulty already in study sections. There is no home for them. They get bounced from study section to study section. I think that is an issue that needs to be dealt with.

Participant: Alaina Fournier, Department of Health and Human Services. I have been working with Dr. Raub for the last 3 months in preparation on this topic for this meeting and future work with this topic, but now I am on a different detail at the NCI (National Cancer Institute) Office of Liaison Activities.

I would like to suggest using the NCI model of getting consumer advocates involved in the scientific process. They have several different programs that have been long, positive, and successful in getting the consumer advocates involved in speaking with the director about the research agenda as well as the peer review process, and just keeping open lines of communication between the two groups.

Participant: I didn't know she was going to talk, but this falls perfectly with what I was going to mention. I have had this conversation with Dr. Raub as well.

I think it is very important to remember that just as our kids have varying—display different aspects that are all different, it is also the way our community is. We have a lot of different viewpoints. It is a wide spectrum.

The point I would like to make about the partnership is, I hope that—and it sounds like from what she just said that the partnership isn't going to be—the determining factor isn't going to be that we have millions of dollars to bring to the table to be able to have our voices heard at the table.

So I think that it would be very important to remember that—to bring all the different advocacy groups to the table so that we all have a place and a voice, because these are all our kids. So I think that is very important to remember.

Participant: Hi, I'm Beth Roy with Social and Scientific Systems. I have been the director of the Pediatric AIDS Clinical Trial Group Operations Center, and I was very interested in hearing about the possibility of looking at a model like that, that I think was successful in bringing together public–private partnerships, the involvement of consumers, and dealing with a very complex disease that was very multidisciplinary. So I just think it would be a good model to look at.

Participant: Jim Moody with SafeMinds. To paraphrase a famous quote, money is the mother's milk of science. When this country has faced a crisis in the past, the Manhattan Project in World War II, the opportunity to land a man on the moon in the 1960s with NASA, and more recently responding to crises like bird flu, SARS, even the recent pet food crisis, there has been a political response which helped drive the scientific response.

So one thing that would I think help would be if this panel of distinguished scientists could urge upon national government the same kind of commitment that has called forth the sense of national urgency at the political level, and that begins at the presidential level, to declare a national emergency, to marshal all of the resources that we can bring to bear on this crisis.

At the present rate of increase, which I think Mark Blaxill said was 10-fold since 1990, in two or three generations every child will be born with autism or related neurological deficit or some sort. These kids now are the canaries in the mine, and there needs to be the strongest possible social response to that. America is full of the top scientists in the world. We can solve any problem if we put our hearts and minds to it. The cost alone of the epidemic would justify that, if not the moral imperative.

Thank you.

Participant: A quick question to all of you here from NIH and the Institutes. How is the so-called funding from the Combat Autism Act recently passed by Congress supposed to be directed or targeted?

Dr. Leshner: I can answer that. It has to be appropriated. It has not been appropriated. It was just authorized.

Participant: A quick comment. I am with the Environmental Protection Agency. My name is Mark Corrales. I just wanted to offer myself as an unofficial contact. I am in the Administrator's office. I am familiar with the various datasets and offices. EPA is a big place. It is sometimes hard to find the right information or people. So I just wanted to offer myself as a contact, if I can help, in the spirit of partnerships, data sharing, and so on.

Dr. Leshner: I think we have gotten to the moment. It feels like I don't want it to end.

One thing that I would like everybody to do, because I think it has been really terrific, and we have been so well behaved and not applauded anyone, we need to take a moment. I think this has been spectacular, and I hope I am right. I would repeat the comment I made fairly glibly before, that is, this is a very important start, and if we don't do something, then shame on us.

Thank you very much.

A

Index of Scientific Opportunities

The following index provides a summary of the many scientific opportunities that were identified by individual workshop speakers and participants. These ideas were not prioritized or debated at the workshop and are not adopted, endorsed, or verified as accurate by the National Academies and as such should not be attributed to the National Academies or the Institute of Medicine. They do, however, represent a rich diversity of possible directions for future research. Each research priority is organized into one of five general categories (Scientific Opportunities for Human Subjects Research, Opportunities Identified to Improve and Enhance Epidemiological Studies, Opportunities to Improve the Understanding of Autism's Pathology, Tools and Infrastructure Needs, and Opportunities for Public–Private Partnerships). In addition, each research priority is referenced to the specific page(s) of the proceedings (Chapter 2) where the concept was discussed and attributed to the associated workshop speaker or participant. Concepts are not listed in order of priority or importance.

Scientific Opportunities for Human Subjects Research	
Update clinical diagnosic criteria for autism spectrum disorder (ASD)	pp. 21, 249 (Swedo)
• Base clinical definition on quantifiable measures rather than qualitative scores if needed	p. 57 (Choi); p. 268 (Singer)
• Establish clinical definitions that allow stratification of ASD subpopulations	pp. 50, 88 (Levitt); p. 89 (Martinez)

Identify and stratify biologically meaningful subpopulations	pp. 31, 88 (Levitt); pp. 38, 234 (Pessah); p. 48 (Herbert); p. 89 (Martinez); pp. 191, 249 (Bernard); p. 240 (Redwood); p. 256 (Falk); p. 267 (Singer); p. 279 (participant)
• Perform genetic sequencing to identify functional variants	p. 32 (Levitt); p. 258 (Goldstein)
• Analyze biological samples, including blood and urine (see also "expand tissue repositories" below in Tools and Infrastructure Needs section)	p. 46 (Herbert); p. 57 (Choi); p. 144 (Redwood); p. 279 (participant)
▪ Perform metabolic profiling	p. 44 (Herbert)
• Establish metabolic profiles of vulnerability and treatability	p. 44 (Herbert); p. 281 (participant)
• Separate individuals with genetic etiology from those with a predominately environmental etiology (that may have some genetic susceptibility) to aid in epigenetic sampling	p. 37 (Pessah); p. 80 (Martinez)
• Perform genetic analysis comparing distinct ASD subpopulations and similar disorders, for example, Asperger's disorder, Fragile X	p. 137 (Swedo)
Identify biomarkers to examine ASD onset, progression, treatment efficacy, metabolic changes, and subpopulations (see also Tools and Infrastructure Needs section)	p. 82 (Lipkin); p. 141 (Leshner); p. 191 (Wilcox); p. 240 (Redwood); p. 242 (Pessah); p. 248 (Blaxill); pp. 268, 271 (Colamarino and Bono); p. 275 (Herbert)
Perform scientifically rigorous analysis of novel and purported treatments	pp. 12, 146, 234, 271 (Bono); pp. 13, 51, 141 (Leshner); pp. 19, 21 (Swedo); p. 90 (Herbert); p. 144 (Blaxill); p. 235 (participant); p. 239 (Noble); p. 269 (Bernard); p. 276 (Zimmerman)

• Use randomized clinical trials	p. 21 (Swedo); p. 243 (Pessah)
• Investigate responses to medication or other forms of treatment individuals are currently undergoing	p. 53 (James); p. 148 (Swedo); p. 154 (Herbert); p. 235 (participant); p. 268 (Redwood); p. 271 (Bono); p. 277 (Zimmerman)
▪ Perform further analysis on "n of 1" studies including serial analyses to capture which biomarkers change with treatment progression	p. 21 (Swedo); p. 45 (Herbert); p. 141 (Insel); p. 256 (Falk); p. 275 (Landrigan)
• Examine individuals' medical records	pp. 12, 271 (Bono); p. 238 (Pessah); p. 263 (Redwood)
• Establish metrics by which to measure onset and recovery	p. 90 (Herbert); pp. 91, 269 (Bernard); p. 234 (Bono); p. 249 (Swedo)
• Perform medical and genetic workups on all known and suspected cases of ASD	p. 276 (Swedo)
• Explore treatment strategies dependent on autism phenotype	p. 263 (Redwood)
• Investigate comorbidities and cases of abnormal onset, for example, *de novo* defects	p. 268 (Bernard); p. 273 (Beaudet)
• Examine ASD throughout an individual's life span, including adolescence and adulthood	p. 223 (Insel); p. 249 (Bernard); p. 251 (Wilcox); p. 270 (Hertz-Picciotto)
Opportunities to Improve and Enhance Epidemiological Studies	
Expand analysis of longitudinal studies utilizing birth cohorts	pp. 73, 229 (Susser); p. 232 (Fombonne); p. 233 (Insel)

• The National Children's Study	pp. 64, 66, 73, 84, 90, 91, 159, 172, 201, 202, 206, 227, 256 (Landrigan, Susser, Schwartz, Alexander, Hertz-Picciotto, Bernard, participant, Schendel, Needham, Pessah, Falk); p. 80 (Martinez); p. 90 (Hertz-Picciotto); p. 233 (Susser)
• MARBLES (Markers of Autism Risk in Babies Learning Early Signs)	p. 201 (Hertz-Picciotto)
Review and assess on going and completed epidemiological studies for gaps and opportunities	pp. 145, 200 (Schwartz); p. 202 (Falk); p. 231 (Schendel)
Collect data and examine time trends to determine if the incidence of ASD has increased	p. 21 (Swedo); p. 142 (Beaudet); p. 176 (Newschaffer); p. 250 (Susser); p. 267 (Singer); pp. 271, 273, 274, 278 (Bono, Blaxill, Susser, Insel)
• Examine incidence data from populations with distinct risk factors	p. 192 (Wilcox); p. 230 (Newschaffer)
• Read retrospective studies that may offer insightful information on time trends	p. 247 (Blaxill); p. 270 (Singer); p. 267 (Newschaffer)
Improve pooling and sharing of data and resources	pp. 182, 188, 202, 232, 270 (Schendel); pp. 189, 245 (Leshner); p. 197 (Bono); pp. 200, 270 (Hertz-Picciotto); p. 245 (Schwartz); p. 258 (Goldstein)
• Standardize criteria used to define cohorts	p. 180 (Newschaffer); p. 182 (Schendel); p. 276 (Swedo)
• Improve and standardize screening criteria	p. 182 (Schendel); p. 236 (Martinez); p. 270 (Newschaffer)

Perform epidemiological analysis using cohorts established through "natural experiments"	p. 73 (Susser); pp. 73, 75, 80, 85, 228, 247, 248, 273 (Susser, Martinez, Schwartz, Fombonne, Blaxill); p. 229 (Fombonne); p. 239 (Swedo); p. 278 (Insel)
• Examine potential similarities and differences in specific ethnic groups, e.g., Hispanic versus non-Hispanic, The Amish	p. 87 (participant); p. 91 (Bernard); pp. 181, 228 (Martinez)
• Examine specific occupational exposures	p. 228 (Schendel)
• Study populations that were exposed prenatally to infectious diseases, toxins from industrial disasters, etc.	p. 229 (Fombonne)
• Compare vaccinated and unvaccinated children	pp. 12, 198, 271 (Bono); p. 198 (Redwood)
• Compare effect of RhoGAM exposure	p. 162 (participant)
Examine potential increases of associated comorbidities	p. 234 (Pessah)
Consider adding genetic analysis to all epidemiological studies examining environmental risks	p. 140 (Noble); p. 145 (Falk)
Consider expansion of analysis of concordance studies in monozygotic and dizygotic twins	p. 54 (Insel); p. 561 (Susser); p. 278 (Fombonne)
Examine cohorts from international countries, especially developing countries	p. 21 (Swedo); pp. 178, 179, 180, 270 (Newschaffer); p. 250 (Falk); p. 278 (Insel)
• Norwegian cohort	p. 82 (Lipkin); p. 82 (Schendel); pp. 116, 201 (Lipkin, Hertz-Picciotto); p. 194 (Wilcox)

Examine cohorts with an elevated risk of autism, e.g., AGRE and Baby Sibs	p. 91 (Bernard); p. 258 (Goldstein)
Coordinate efforts with the toxicology field	p. 172 (Hertz-Picciotto)
Opportunities to Improve the Understanding of Autism's Pathology and Etiology	
Examine potential impact of convergence of multiple types of stressors	p. 44 (Herbert); p. 115 (Lipkin); p. 131 (Slotkin)
Elucidate the potential role of immune system and immunological susceptibilities	pp. 12, 271 (Bono); pp. 38, 242 (Pessah); pp. 259, 269 (Goldstein); p. 272 (Noble); p. 277 (Cohen)
• Expand current efforts in microbiomics	p. 118 (Lipkin); p. 278 (Insel)
• Investigate the role of biopathogens on organ systems, including nervous and digestive	p. 251 (Bono)
Focus on toxicology	
• Examine effects of subtoxic exposure levels on different organ systems	pp. 38, 276 (Pessah)
• Collect data on neurodevelopmental toxicity of major priority pollutants, including toxokinetic study	p. 66 (Landrigan); p. 205 (Needham)
• Examine the potential impact of heavy metals and cosmetics	p. 12 (Bono); p. 87 (participant); p. 240 (Hertz-Picciotto)
▪ Perform toxokinetic studies	pp. 250, 271 (Bono),
▪ Develop distribution and excretion profiles	p. 239 (Noble); p. 257 (Falk)
• Examine potential effects of toxicant synergies	p. 34 (Pessah); p. 244 (Noble)
Examine the potential role of oxidative stress	p. 237 (Needham)
Investigate the potential role of metabolism and metabolic responses/dysfunctions	pp. 44, 128 (Herbert)

• Explore impact on modulating brain function and vice versa	p. 281 (participant)
Generate gene expression profiles, both time and topology	p. 32 (Levitt); p. 258 (Goldstein)
Examine the potential impact of genetic mutations and environmental toxicants on the development and maintenance of neuronal circuitry	pp. 29, 32 (Levitt); p. 41 (Herbert); p. 98 (Beaudet); p. 241 (Insel); p. 275 (Pessah)
Examine autism pathology	p. 41 (Herbert); p. 258 (Goldstein)
• Explore changes over time in nervous system function, neurotransmitter profiles, and neuronal circuitry	p. 21 (Swedo)
• Examine impact on germ-line cells	p. 278 (Fombonne)
Expand development and use of imaging as a mechanism to examine progression of autism pathology	pp. 41, 45 (Herbert); p. 276 (Spence); p. 277 (Cohen)
Develop chelators	p. 271 (Hertz-Picciotto)
• Formulate chelators that can cross the blood-brain barrier	p. 12 (Bono)
• Examine efficacy of glutathione as a chelator	p. 87 (participant)
• Investigate impact of detoxification of organ systems and its possible impact on inflammation and gastrointestinal problems	p. 271 (Bono)
Expand efforts to perform epigenetic analysis	p. 80 (Martinez); pp. 98, 273 (Beaudet); p. 241 (Insel); p. 250 (Bono)
Expand development and use of models to study ASD	pp. 30, 32 (Levitt); p. 243 (Pessah); p. 252 (Martinez); p. 272 (Noble)
• Develop high-throughput models, e.g., cell culture methods	p. 112 (Slotkin)

• Create animal models that reflect pathology in specific organ systems	p. 243 (Pessah)
• Establish primary cell lines from autistic individuals	p. 276 (Pessah)
• Study related disorders, e.g., cystic fibrosis and Fragile X	p. 237 (Swedo); p. 277 (Coetzee)
Tools and Infrastructure Needs	
Expand tissue repositories	p. 150 (Insel); pp. 150, 250, 265, (Schwartz); pp. 171, 224 (Hertz-Picciotto, Falk, Schendel); p. 268 (Needham); p. 271 (James); p. 279 (participant)
• Require standardized sample procurement and storage	p. 215 (Walt); p. 227 (Pessah); p. 269 (Goldstein)
Establish programs to expand the number of investigators trained in environmental biology	p. 66 (Landrigan); pp. 221, 265 (Schwartz); p. 221 (Goldstein)
Establish incentive programs to attract individuals from other fields to study ASD	p. 270 (Newschaffer)
Develop biomarkers of exposure, susceptibility, state, outcome; biomarkers studies that are replicable; biomarkers to spot the effects of environmental neurotoxicants	p. 82 (Lipkin); p. 134 (Insel); p. 172 (Hertz-Picciotto); p. 240 (Redwood); p. 247 (Blaxill); p. 250 (Newschaffer)
Invest in the development of improved tools to analyze and multiplex data, including environmental sampling, biological samples, and potential metabolic changes	p. 216 (Walt)
Develop a large clinical trial network	p. 265 (Schwartz); p. 276 (Swedo)
• Expand autism patient and family registries	p. 54 (Insel); p. 73 (Susser); pp. 235, 236 (participants); p. 237 (Beaudet); p. 269 (Goldstein); p. 276 (Swedo)

Establish and expand autism centers of excellence to study children's environmental health	p. 66 (Landrigan)
Expand resources to identify potential environmental factors	p. 82 (Lipkin); p. 208 (Needham); pp. 215, 221 (Walt); p. 238 (Pessah); p. 241 (Hertz-Picciotto); p. 252 (Martinez)
• Develop and implement environmental sensor and biosensor technology	
▪ Take advantage of continuous sampling	p. 211 (Walt)
• Use strategies that prioritize potential environmental stressors or use an unbiased strategy	p. 207 (Needham); p. 215 (Walt); pp. 268, 271 (Singer)
• Expand use of technologies that allow high-throughput analysis	p. 205 (Slotkin)
▪ For example, HPLC, mass spectroscopy, two-dimensional gas chromatography, "lab-on-a-chip"/microfluidics	p. 207 (Needham); pp. 213, 214 (Walt)
Opportunities for Public-Private Partnerships	
Increase coordination and integration of each stakeholder's ongoing and planned efforts	p. 246 (Blaxill); p. 269 (Redwood)
Establish enhanced methods of coordination and data-sharing policies for public-private partnerships	p. 188 (Schendel); p. 259 (Goldstein); p. 269 (Schwartz)
• Establish collaborations with small, innovative biotech companies	p. 221 (Walt); p. 277 (Coetzee)
Improve public engagement in the development of research priorities	p. 260 (Insel); pp. 262, 269 (Redwood); p. 267 (Leshner)

B

Workshop Agenda
Autism and the Environment: Challenges and Opportunities for Research

Wednesday, April 18, 2007
Keck Building, Room 100
500 Fifth Street, N.W.
Washington, DC 20001

Workshop Objectives

- Discuss the most promising scientific opportunities for improving the understanding of potential environmental factors in autism.
- Discuss what scientific tools and technologies are available, what interdisciplinary research approaches are needed, and what further infrastructure investments will be necessary in the short and long term to be able to explore potential relationships between autism and environmental factors.
- Explore potential partnerships needed to support and conduct autism research.

8:00 a.m. Welcome, Introductions, and Workshop Objectives

ALAN LESHNER
Workshop and Forum Chair
Chief Executive Officer, AAAS
Executive Publisher, *Science*

8:05 a.m. Charge to Workshop Participants

WILLIAM RAUB
Science Advisor to the Secretary
Department of Health and Human Services

8:15 a.m. Perspectives of the Advocacy Community

LAURA BONO
Workshop Planning Committee Member
Board Member
National Autism Association

SESSION I: AUTISM—THE CLINICAL PROBLEM: "WHAT DO WE KNOW? WHAT DO WE NEED?"

Session Objective: Describe the problem and discuss how environmental factors may impact a developmental disorder like autism. Identify what standards of evidence are needed to move forward.

SARAH SPENCE, *Session Chair*
Staff Clinician
Pediatrics and Developmental Neuropsychiatry Branch
National Institute of Mental Health, NIH

8:25 a.m. Clinical Overview: How Can the Clinical Manifestations of Autism Shed Light on Potential Environmental Etiologies?

SUSAN SWEDO
Senior Investigator
Pediatrics and Developmental Neuropsychiatry Branch
National Institute of Mental Health, NIH

8:45 a.m. Genes and the Environment: How May Genetics Be Used to Inform Research Searching for Potential Environmental Triggers?

PATRICK LEVITT
Director
Vanderbilt Kennedy Center for Research on Human Development, Vanderbilt University

9:05 a.m. How May Environmental Factors Impact Potential Mechanisms in Humans?

ISAAC PESSAH
Director
Children's Center for Environmental Health and Disease Prevention
University of California–Davis M.I.N.D. Institute

9:25 a.m. Defining Autism: Biomarkers and Other Research Tools

MARTHA HERBERT
Assistant Professor of Neurology
Harvard Medical School

9:45 a.m. Discussion

SARAH SPENCE, *Session Chair*
Staff Clinician
Pediatrics and Developmental Neuropsychiatry Branch
National Institute of Mental Health, NIH

10:20 a.m. BREAK

SESSION II: LESSONS LEARNED FROM OTHER DISORDERS: "STANDARDS OF EVIDENCE"

Session Objective: Explore how the autism field may employ approaches and strategies used by other fields. What has been learned from research that has examined environmental exposure effects on other disorders?

DAVID SCHWARTZ, *Session Chair*
Director
National Institute of Environmental Health Sciences, NIH

10:35 a.m. Environmental Toxicants and Neurodevelopment

PHILIP LANDRIGAN
Chair of Community and Preventive Medicine
Mt. Sinai School of Medicine

10:55 a.m. Prenatal Starvation and Schizophrenia

EZRA SUSSER
Gelman Professor and Chair of Epidemiology
Mailman School of Public Health
Professor of Psychiatry
Columbia University and New York State Psychiatric Institute

11:15 a.m. Asthma

FERNANDO MARTINEZ
Swift-McNear Professor of Pediatrics
Director, Arizona Respiratory Center
University of Arizona

11:35 a.m. Discussion

DAVID SCHWARTZ, *Session Chair*
Director
National Institute of Environmental Health Sciences, NIH

12:10 p.m. LUNCH

SESSION III: ENVIRONMENT AND BIOLOGY I: WHAT ARE THE TOOLS FOR AUTISM— WHAT DO WE HAVE? WHAT DO WE NEED?

Session Objective: Review how environmental factors can impact fundamental biological processes. Examine the resources available, and needed, to examine susceptibility to environmental agents.

PATRICK LEVITT, *Session Chair*
Director
Vanderbilt Kennedy Center for Research on Human Development, Vanderbilt University

1:00 p.m. How May Environmental Factors Impact Potential Molecular and Epigenetic Mechanisms?

ARTHUR BEAUDET
Professor and Chair
Department of Molecular and Human Genetics
Baylor College of Medicine

1:20 p.m. How May Environmental Factors Impact Potential Cell-Based Mechanisms?

MARK NOBLE
Professor of Genetics
University of Rochester Medical Center

1:40 p.m. How May Animal Models Be Used to Examine Potential Environmental-Based Mechanisms?

THEODORE SLOTKIN
Professor of Pharmacology and Cancer Biology
Duke University Medical Center

2:00 p.m. BREAK

2:20 p.m. Autism, Infection, and Immunity: What Are the Potential Causative Environmental Factors and How Can They Be Identified and Prioritized?

W. IAN LIPKIN
Director, Columbia Center for Infection and Immunity
Mailman School of Public Health of Columbia University, and
Scientific Director, Northeast Biodefense Center

2:40 p.m. Environmental Factors and Oxidative Stress: How May Oxidative Stress Impact the Biology of Autism? What Factors May Be Causing This Outcome?

S. JILL JAMES
Professor of Pediatrics
University of Arkansas for Medical Sciences

3:00 p.m. Discussion

PATRICK LEVITT, *Session Chair*
Director
Vanderbilt Kennedy Center for Research on Human Development, Vanderbilt University

SESSION IV: NEW APPROACHES AND DISCUSSION WITH WORKSHOP ATTENDEES

4:20 p.m. Discussion with Meeting Participants and Audience

ALAN LESHNER, *Moderator*
Workshop and Forum Chair
Chief Executive Officer, AAAS
Executive Publisher, *Science*

5:30 p.m. ADJOURN

Thursday, April 19, 2007
Keck Building, Room 100
500 Fifth Street, N.W.
Washington, DC 20001

SESSION V: ENVIRONMENTAL EPIDEMIOLOGY—UTILIZING POPULATION-BASED STUDIES TO ISOLATE THE ENVIRONMENTAL CAUSES OF AUTISM

Session Objective: Discuss and identify what resources are available and what is needed to help frame future directions for environmental epidemiology studies.

HENRY FALK, *Session Chair*
Director
Coordinating Center for Environmental Health and Injury Prevention, Centers for Disease Control and Prevention

8:00 a.m. Environmental Epidemiology Studies: New Techniques and Technologies to Use Epidemiology to Find Environmental Triggers

IRVA HERTZ-PICCIOTTO
Professor of Epidemiology and Preventive Medicine
University of California–Davis

8:20 a.m. Environmental Exposures in Autism: International Studies

CRAIG NEWSCHAFFER
Professor and Chairman of Epidemiology and Biostatistics
Drexel University

8:40 a.m. Environmental Epidemiology Studies: CADDRE

DIANA SCHENDEL
National Center on Birth Defects and Developmental Disabilities
Centers for Disease Control and Prevention

9:00 a.m. Prenatal and Perinatal Exposures

ALLEN WILCOX
Chief
Epidemiology Branch
National Institute of Environmental Health Sciences, NIH

9:20 a.m. Discussion

HENRY FALK, *Session Chair*
Director
Coordinating Center for Environmental Health and Injury Prevention, Centers for Disease Control and Prevention

SESSION VI: TECHNOLOGY AND INFRASTRUCTURE NEEDS FOR FUTURE RESEARCH

Session Objective: Discuss and identify what tools are currently available to assess environmental exposure, and what additional scientific tools and technologies are needed in the short and long term.

HENRY FALK, *Session Chair*
Director
Coordinating Center for Environmental Health and Injury Prevention, Centers for Disease Control and Prevention

9:50 a.m. CDC Environmental Health Lab—Body Burden Measures

LARRY NEEDHAM
Chief
Organic Analytical Toxicology Branch
National Center for Environmental Health, CDC

10:10 a.m. Personalized Environmental Sensors

DAVID WALT
Professor of Chemistry
Tufts University

10:30 a.m. Discussion

HENRY FALK, *Session Chair*
Director
Coordinating Center for Environmental Health and Injury Prevention, Centers for Disease Control and Prevention

SESSION VII: FUTURE RESEARCH DIRECTIONS—DISCUSSION WITH WORKSHOP SPEAKERS

Session Objective: Discuss what a research agenda for autism and the environment might look like.

11:00 a.m. FERNANDO MARTINEZ
Swift-McNear Professor of Pediatrics
Director, Arizona Respiratory Center
University of Arizona

ISAAC PESSAH
Director, Children's Center for Environmental Health and Disease Prevention
University of California–Davis M.I.N.D. Institute

12:30 p.m. LUNCH

SESSION VIII: PUBLIC–PRIVATE PARTNERSHIPS

Session Objective: Identify the unique strengths that the public and private sectors provide to the autism research. Discuss how each of these sectors can most effectively complement each other's efforts.

ALAN LESHNER, *Moderator*
Workshop and Forum Chair
Chief Executive Officer, AAAS
Executive Publisher, *Science*

1:00 p.m. Panel Discussion

SALLIE BERNARD
Board Member, Autism Speaks
Cofounder, SafeMinds

HENRY FALK
Director
Coordinating Center for Environmental Health and Injury Prevention, Centers for Disease Control and Prevention

GARY GOLDSTEIN
Chair, Autism Speaks Scientific Affairs Committee
President and Chief Executive Officer, Kennedy Krieger Institute

TOM INSEL
Director
National Institute of Mental Health, NIH

LYN REDWOOD
Board Member and Science Committee Chair
National Autism Association

DAVID SCHWARTZ
Director
National Institute of Environmental Health Sciences, NIH

1:45 p.m. General Panel Discussion

ALAN LESHNER, *Moderator*
Workshop and Forum Chair
Chief Executive Officer, AAAS
Executive Publisher, *Science*

SESSION IX: DISCUSSION WITH MEETING PARTICIPANTS AND AUDIENCE

2:15 p.m. ALAN LESHNER, *Moderator*
Workshop and Forum Chair
Chief Executive Officer, AAAS
Executive Publisher, *Science*

3:15 p.m. ADJOURN

C

Registered Workshop Participants

Gregory Allsberry
Foundation for Mercury Injured Children

Karla Allsberry
Foundation for Mercury Injured Children

Melissa Aubrey
Sheppard Pratt

Peter Bell
Autism Speaks

Elizabeth Blackburn

Douglas Boenning
Children's National Medical Center

Scott Bono
National Autism Association

Brandon Boxler
U.S. Department of Justice

Ann Brasher
National Autism Association

Kathleen Brennan

Kristin Butterfield
American Academy of Physician Assistants

Keely Cheslack-Postava
Johns Hopkins School of Public Health

Margaret Chu
U.S. Environmental Protection Agency

Lujene Clark
NoMercury

Beth Clay
BC&A International, LLC

Mark Corrales
U.S. Environmental Protection Agency

Alison Davis

Kelli Ann Davis

Flauren De Souza

Vicky Debold

Gayle DeLong

Richard Deth
Northeastern University

Patrick Dollard
The Center for Discovery

Jim Donnelly
Fuzz Foundation for Autism

Heather Elias

Richard Emerson

Albert Enayati
SafeMind

Rebecca Estepp
Talk About Curing Autism

Joseph Evall
Orrick

Pat Fasick
Annapolis Children's Therapy Center

Debra Fellner

Alaina Fournier
U.S. Department of Health and Human Services

Renee Gardner
Johns Hopkins Bloomberg School of Public Health

Sydney Gary
Cold Spring Harbor Laboratory

Lisa Gilotty
National Institute of Mental Health

Chelsey Goddard

Harold Grams

Rebecca Grant-Widen
National Autism Association

George Hajduczok

Alycia Halladay
Autism Speaks

Tamar P. Halpern

James Hanson
National Institute of Child Health and Human Development

Wendy Harnisher
Capital District Biomedical Support Group for Autism Spectrum Disorders

Martha M. Harris

Susanne Harris

Laurie Henrikson
The Aerospace Corporation

Deborah Hirtz
National Institute of Neurological Disorders and Stroke

Mady Hornig
Mailman School of Public Health

Valerie Hu
The George Washington University Medical Center

Erica Johnson

Linda Kahan
Food and Drug Administration

Alice Kau
National Institute of Child Health and Human Development

Sharla Khargi
SCO Family of Services

Dawn Koplos

Robert Krakow

John Kucera

Valérie La Traverse

Rebecca Lane
National Institute of Mental Health

Scott Laster
National Autism Association

Paul Law
Kennedy Krieger Institute

Cindy Lawler
National Institute of Environmental Health Sciences

Aaron Levin
Psychiatric News

Rebecca Loveszy
Family Unity International, Inc. and Music Therapy Services of New York

Richard Loveszy
Family Unity International, Inc. and Music Therapy Services of New York

Joseph Lowe
U.S. Court of Federal Claims

Bobbie Manning
A-CHAMP

W. John Martin

Alicia Mastronardi
National Institute of Allergy and Infectious Diseases

June McCullough
San Jacinto Unified School District

Stephanie McFadden

Mary McKenna
University of Maryland School of Medicine

Rita McWilliams
UMD-NJ/RWJ Medical School

Elieana Mihai

Claudia Miller
University of Texas Health Science Center

Barbara Mulach
National Institute of Child Health and Human Development

Amy Nevel
HHS/ASPE/Office of Science and Data Policy

Jennifer Nyland
Johns Hopkins Bloomberg School of Public Health

Molly Oliveri
National Institute of Mental Health

Dan Olmsted
United Press International

Raymond Palmer
University of Texas Health Science Center–San Antonio

Lynne Parsons-Heilbrun

Aimee Peer

Anabel Perez

Becky Peters

Thomas Powers
Williams Love O'Leary Craine & Powers

Rajendram Rajnarayanan

Marnina Reed
National Institutes of Health

Catherine Rice
Centers for Disease Control and Prevention

Rebecca Rienzi

Robert Rinicella

Daphne Robinson
National Institute of Neurological Disorders and Stroke

Stephanie Ross
ASPH based at U.S. Environmental Protection Agency

Beth Roy
Social & Scientific Systems

Leslie Rutherford

Pam Schwingl
Social & Scientific Systems

Jeff Sell
Autism Society of America

Janet Sheehan

Rita Shreffler
National Autism Association

Beth-Anne Sieber
National Institute of Mental Health

Thomas Sinks
Centers for Disease Control and Prevention

Kirsten Thistle

Daniel Thomasch
Orrick

Louise Tiranoff

Victor Ty

Ann Wagner
National Institute of Mental Health

Tatjana Walker
University of Texas Health Science Center–San Antonio

Gemma Weiblinger
National Institute of Mental Health

Julia Whiting

J. Kenneth Wickiser
The Rockefeller University

Margaret Williams

Nikki Withrow

Bob Wright
Autism Speaks

Suzanne Wright
Autism Speaks

Gerardine Wurzburg
State of the Art

Marshalyn Yeargin-Allsopp
Centers for Disease Control and Prevention

Kathy Young

Lisa Zbar

Renjian Zhao

Paul B. Zuydhoek

D

Biographic Sketches of Workshop Planning Committee, Forum Members, Invited Speakers, and Staff

WORKSHOP PLANNING COMMITTEE

Alan I. Leshner, Ph.D. (*Workshop Chair, Neuroscience Forum, Chair*), is chief executive officer (CEO) of the American Association for the Advancement of Science (AAAS) and executive publisher of its journal, *Science*. Previously Dr. Leshner had been director of the National Institute on Drug Abuse (NIDA) at the National Institutes of Health (NIH) and deputy director and acting director of the National Institute of Mental Health (NIMH). Before that, he held a variety of senior positions at the National Science Foundation (NSF). Dr. Leshner began his career at Bucknell University, where he was professor of psychology. Dr. Leshner is an elected member of the Institute of Medicine (IOM) of the National Academies, and a fellow of AAAS, the National Academy of Public Administration, and the American Academy of Arts and Sciences. He was appointed by the U.S. President to the National Science Board, and is a member of the Advisory Committee to the Director of NIH. He received an A.B. in psychology from Franklin and Marshall College and an M.S. and a Ph.D. in physiological psychology from Rutgers University. Dr. Leshner also holds honorary Doctor of Science degrees from Franklin and Marshall College and Pavlov Medical University in St. Petersburg, Russia.

Duane Alexander, M.D., was named director of the National Institute of Child Health and Human Development (NICHD) in 1986, after serving as the Institute's acting director. Much of his career has been with NICHD. After receiving his undergraduate degree from Pennsylvania State University, Dr. Alexander earned his M.D. from Johns

Hopkins University School of Medicine. Following his internship and residency at the Department of Pediatrics at Johns Hopkins Hospital, he joined NICHD in 1968 as a clinical associate in the Children's Diagnostic and Study Branch. Following his tenure with the branch, Dr. Alexander returned to Johns Hopkins as a fellow in pediatrics (developmental disabilities) at the John F. Kennedy Institute for Habilitation of the Mentally and Physically Handicapped Child. He returned to NICHD in 1971, when he became assistant to the scientific director and directed the NICHD National Amniocentesis Study. He is a diplomate of the American Board of Pediatrics and a member of the American Academy of Pediatrics, the American Pediatric Society, and the Society for Develop-mental Pediatrics. For more than a decade, he also served as the U.S. observer on the Steering Committee on Bioethics for the Council of Europe. As an officer in the Public Health Service (PHS), Dr. Alexander has received numerous PHS awards, including a Commendation Medal in 1970, a Meritorious Service Medal, and a Special Recognition Award in 1985. He also received the Surgeon General's Exemplary Service Medal in 1990, and the Surgeon General's Medallion in 1993 and 2002. In addition, Dr. Alexander is the author of numerous articles and book chapters, most of which relate to his research in developmental disabilities.

Mark Blaxill is the father of a daughter diagnosed with autism and vice president of SafeMinds. He spent most of his professional career at the Boston Consulting Group (BCG), where he was a senior vice president. While at BCG he was the leader of the firm's Strategy Practice and led firm initiatives in the area of globalization, open source software, intellectual property, and network analysis. He has wide industry experience, including client assignments in information services, pharmaceuticals, consumer electronics, and retailing. He has worked on a wide range of business problems for CEOs and heads of strategy of Fortune 100 and Dow Jones Index companies. He is writing a book on the subject of intellectual property strategies for business and launching a new business venture. He is a named inventor on BCG's first patent application. He holds an M.B.A. from Harvard Business School and a bachelor's degree in international affairs from Princeton University. He is also the author of several publications on autism, including "What's Going On? The Question of Time Trends in Autism" (*Public Health Reports*, 2004); "Reduced Mercury Levels in First Baby Haircuts of Autistic Children" (*International Journal of Toxicology*, 2003); and "Thimerosal and Autism? A Plausible Hypothesis That Should Not Be Dismissed" (*Medical Hypotheses,* 2004). He has been a frequent speaker on autism-related issues, including conference presentations for Neurotoxicology (2006), Defeat Autism Now! (2001, 2006), Autism One (2004, 2005, and 2006), National Autism Association (2005), National Institute of Environmental Health Sciences (NIEHS) (2005), and the IOM Immunization Safety Review (2001).

Laura Bono is a board member, cofounder, and chair emeritus of the National Autism Association (NAA). Along with NIEHS and SafeMinds, she helped to plan and execute the Environmental Factors in Neurodevelopmental Disorders Symposium. She graduated from the University of South Carolina with a B.S. in journalism with a minor in marketing. She currently serves as director of marketing for the Parent Institute and has more than 25 years of business experience in marketing. The youngest of her three children has been diagnosed with autism spectrum disorder.

Sophia Colamarino, Ph.D., is vice president of research at Autism Speaks. Among her duties, Dr. Colamarino manages and oversees Autism Speaks' Biology Portfolio and new High Risk/High Impact program. After 16 years of research experience, she joined Cure Autism Now (CAN) in November 2004 as Science Director to oversee the science program in association with the CAN Scientific Review Council. She graduated with dual degrees in biological sciences and psychology from Stanford University. She received her Ph.D. in neurosciences from University of California–San Francisco (UCSF), where she studied brain development with distinguished neuroscientist Marc Tessier-Lavigne, Ph.D. After receiving her Ph.D., she conducted research on the genetic disorder Kallmann Syndrome at the Telethon Institute for Genetics and Medicine in Milan, Italy, led by human geneticist Andrea Ballabio, M.D. She then returned to the United States to work at the Salk Institute in La Jolla, CA, studying adult neural stem cells and brain regeneration in the laboratory of stem cell pioneer Fred H. Gage, Ph.D. Dr. Colamarino's research career has included publications in journals such as *Cell* and *Nature*. During her tenure at CAN, Dr. Colamarino oversaw a large growth in the science program, expanding the CAN research portfolio from 11 grants in 2004 to 39 in 2006, and developed several important autism initiatives, including the Neuropathology Workgroup, a collaborative effort to understand the cellular and molecular basis of brain enlargement in autism, the first Environmental Innovator Award, and research summit meetings on immunology and neuroimaging.

Eric Fombonne, M.D., is the head of the Division of Child Psychiatry at McGill and director of the Department of Psychiatry at the Montreal Children's Hospital, where he has expanded autism services. He worked at INSERM in France and at the London Institute of Psychiatry in England, and he is now holder of a Canada Research Chair. He has been involved in numerous epidemiological studies of autism and is considered to be a leading authority on this topic, and also on the putative links between autism and immunization. He has also been involved in the development of assessment tools for clinical and research purposes, in family and genetic studies of autism, and in outcome studies. He has a long track record of scientific/research leadership—including serving as a consultant for the National Academies, the Centers for Diseases Control and Prevention (CDC), the American Academy of Pediatrics, the MRC (United Kingdom), and the M.I.N.D. Institute (University of California–Davis)—on research matters related to autism. He has been associate editor of the *Journal of Autism and Developmental Disorders* since 1994 and is on the editorial board of several other scientific journals. He is on the board of several family associations, with which he has worked closely over the years.

Steve Hyman, M.D., is provost of Harvard University and professor of neurobiology at Harvard Medical School. From 1996 to 2001, he served as director of the National Institute of Mental Health (NIMH). Earlier, Dr. Hyman was professor of psychiatry at Harvard Medical School, director of psychiatry research at Massachusetts General Hospital (MGH), and the first faculty director of Harvard University's Mind, Brain, and Behavior Initiative. In the laboratory he studied the regulation of gene expression by neurotransmitters, especially dopamine, and by drugs that influence dopamine systems. This research was aimed at understanding addiction and the action of therapeutic psychotropic drugs. Dr. Hyman is a member of the IOM, a fellow of the American Academy of Arts and Sciences, and a fellow of the American College of Neuropsychopharmacology. He is editor-in-chief of the *Annual Review of Neuroscience.* He has received awards for public service from the U.S. government and from patient advocacy groups such as the National Alliance for the Mentally Ill and the National Mental Health Association. Dr. Hyman received his B.A. from Yale College and an M.A. from the University of Cambridge, which he attended as a Mellon fellow studying the history and philosophy of science. He earned his M.D. from Harvard Medical School.

Judy Illes, Ph.D., is professor of neurology and Canada Research Chair in Neuroethics, National Core for Neuroethics at the University of British Columbia. Dr. Illes received her doctorate in hearing and speech sciences from Stanford University in 1987, with a specialization in experimental neuropsychology. Dr. Illes returned to Stanford University in 1991 to help build the research enterprise in imaging sciences in the Department of Radiology. She also cofounded the Stanford Brain Research Center (now the Neuroscience Institute at Stanford), and served as its first executive director between 1998 and 2001. Today, Dr. Illes directs a strong research team devoted to neuroethics, and issues specifically at the intersection of medical imaging and biomedical ethics. Dr. Illes has written numerous books and edited volumes and articles. She is the author of *The Strategic Grant Seeker: Conceptualizing Fundable Research in the Brain and Behavioral Sciences* (1999, LEA Publishers, NJ), special guest editor of "Emerging Ethical Challenges in MR Imaging," *Topics of Magnetic Resonance Imaging* (2002), and "Ethical Challenges in Advanced Neuroimaging," *Brain and Cognition* (2002). Her latest book, *Neuroethics: Defining the Issues in Theory, Practice and Policy*, was published by Oxford University Press in January 2006. Dr. Illes is co-chair of the Committee on Women in Neuroscience for the Society for Neuroscience; a member of the Internal Advisory Board for the Institute of Neurosciences, Mental Health and Addiction of the Canadian Institutes of Health Research; and a member of the Dana Alliance for Brain Initiatives.

Thomas R. Insel, M.D., is director of the National Institute of Mental Health. He graduated from Boston University, where he received a B.A. from the College of Liberal Arts and an M.D. He did his internship at Berkshire Medical Center, Pittsfield, MA, and his residency at the Langley Porter Neuropsychiatric Institute at UCSF. Dr. Insel joined NIMH in 1979, where he served in various scientific research positions until 1994, when he went to Emory University as professor, Department of Psychiatry, Emory University School of Medicine, and director of the Yerkes Regional Primate Research Center. As director of Yerkes, Dr. Insel built one of the nation's leading HIV vaccine research programs. He also served as the founding director of the Center for Behavioral Neuroscience, a Science and Technology Center funded by NSF to develop an interdisciplinary consortium for research and education at eight Atlanta colleges and universities. Dr. Insel's scientific interests have ranged from clinical studies of obsessive-compulsive disorder to explorations of the molecular basis of social behaviors in rodents and nonhuman primates. His research on oxytocin and affiliative behaviors helped to launch the field of social neuroscience. He oversees NIMH's $1.4 billion research budget, which provides support to investigators at universities in the areas of basic science; clinical research, including large-scale trials of new treatments; and studies on the organization and delivery of mental health services.

David A. Schwartz, M.D., is the director of the National Institute of Environmental Health Sciences (NIEHS) and the National Toxicology Program at NIH. Prior to this appointment, he served on the faculty at the University of Iowa (1988–2000) and Duke University (2000–2005). At Duke University, Dr. Schwartz served as the director of the Division of Pulmonary, Allergy, and Critical Care Medicine and vice chair for research in the Department of Medicine. In this capacity, Dr. Schwartz established three NIH Centers: a center focusing on Environmental Genomics, a Program Project in Environmental Asthma, and an Environmental Health Sciences Research Center. Dr. Schwartz has focused his research on the genetic and biological determinants of environmental and occupational lung disease. These research efforts have provided new insights into the pathophysiology and biology of asbestos-induced lung disease, interstitial lung disease, environmental airway disease, and innate immunity. This research has identified endotoxin or lipopolysaccharide as an important cause of airway disease among those exposed to organic dusts, and determined that a specific mutation in the Toll-4 gene is associated with a diminished airway response to inhaled LPS in humans. Recent work is focusing on the genes that regulate the innate immune response in humans, genes involved in the fibroproliferative response in the lung, and the genetic regulation of environmental asthma. Dr. Schwartz has served on numerous study sections and editorial boards, is a member of the American Society for Clinical Investigation and the Association of American Physicians, and was awarded the Scientific Accomplishment Award from the American Thoracic Society in 2003.

Alison Tepper Singer is executive vice president, awareness and communications, for Autism Speaks. Ms. Singer has been with the foundation since its launch in March 2005 and is a staff-liaison to the board of directors, in addition to overseeing the strategic communications and development of the growing organization. She served as interim CEO of the organization from March through July 2005. Prior to joining Autism Speaks, Ms. Singer spent 14 years at CNBC and NBC in a variety of positions. From 1994 to 1999, she served as vice president of programming in NBC's Cable and Business Development division. Most recently at CNBC, in her role as special projects producer, Ms. Singer produced the award-winning series "Autism: Paying the Price." She has a B.A. in economics from Yale University and an M.B.A. from Harvard Business School. Alison has a daughter and an older brother with autism.

Susan Swedo, M.D., received her B.A. from Augustana College and her M.D. from Southern Illinois University. Shortly after completing a residency in pediatrics at Northwestern University, Dr. Swedo was named chief of the Division of Adolescent Medicine at the university. The following year, she moved to Washington, DC, and became a senior staff fellow in the Child Psychiatry Branch, NIMH. Dr. Swedo was granted tenure in 1992, became head of the Section on Behavioral Pediatrics in 1994, and chief of the Pediatrics and Developmental Neuropsychiatry Branch in 1998. She also served as the acting scientific director for NIMH from 1995 through 1998. Dr. Swedo recently received the Joel Elkes International Research Award from the American College of Neuropsychopharmacology. Her laboratory studies include childhood-onset obsessive-compulsive disorder and related disorders, including Tourette syndrome and Sydenham chorea.

Christian G. Zimmerman, M.D., FACS, M.B.A., is chair and founder of the Idaho Neurological Institute (INI), adjunct professor of psychology at Boise State University, past CEO of Neuroscience Associates, and former board member for the Idaho State Board of Health and Welfare. Dr. Zimmerman established the INI research facility to focus on nervous system injury, repair, and neuroplasticity; leads its various interdisciplinary research teams; and is coprofessor for biology and cognitive neuroscience research students trained at the facility. Research projects include a 20-year longitudinal study of traumatic brain injury; investigations of spinal injury, stroke, aneurysms, arterial thrombolytic therapy intervention, neuropathology, central nervous system (CNS) tumors, sleep disorders, deep-brain stimulation, and movement disorders; and five Telemedicine and Advanced Technologies Research Center (TATRC) grants. In his role as INI chair, he has facilitated numerous symposia and workshops to provide educational opportunities for medical professionals and the public. Additionally, he chairs prevention programs for Idaho's youth such as Think First. Dr. Zimmerman is a diplomate of the American Board of Neurological Surgery and Pain Management and a fellow of the American College of Surgeons and Physician Executives. He received his M.B.A. from Auburn University, and his M.D. from the University of Maryland.

FORUM MEMBERS

Alan I. Leshner, Ph.D. (*Chair*), biography in Workshop Planning Committee.

Huda Akil, Ph.D., is the Gardner Quarton Distinguished University Professor of Neuroscience and Psychiatry at the University of Michigan, and the codirector of the Molecular and Behavioral Neuroscience Institute. Dr. Akil has made seminal contributions to the understanding of the neurobiology of emotions, including pain, anxiety, depression, and substance abuse. Early on, she focused on the role of the endorphins and their receptors in pain and stress responsiveness. Dr. Akil's scientific contributions have been recognized with numerous honors and awards. These include the Pacesetter Award from NIDA in 1993, and with Dr. Stanley Watson, the Pasarow Award for Neuroscience Research in 1994. In 1998, she received the Sachar Award from Columbia University and the Bristol Myers Squibb Unrestricted Research Funds Award. Dr. Akil is past president of the American College of Neuropsychopharmacology (1998) and past president of the Society for Neuroscience (2004), the largest neuroscience organization in the world. She was elected as a fellow of AAAS in 2000. In 1994, she was elected to be a member of the IOM and is currently a member of its Council. In 2004, she was elected to the American Academy of Arts and Sciences. Dr. Akil received her Ph.D. from the University of California, Los Angeles.

Marc Barlow joined the strategic marketing group in GE Healthcare as leader of the neuroscience area in 2005. In this role he is responsible for the development and delivery of disease area strategies for CNS. Before joining GE Mr. Barlow was the marketing director of Sanofi-Aventis in the United Kingdom. Prior to this he held a number of senior sales and marketing positions within the pharmaceutical industry, both domestically in the United Kingdom and internationally based out of the United States and Switzerland. A large amount of Mr. Barlow's experience has been in the neuroscience area, in particular in epilepsy, Alzheimer's disease, and stroke. Mr. Barlow graduated from the University of Wolverhampton with a focus in biological sciences and the Chartered Institute of Marketing with a diploma in marketing studies.

Dennis W. Choi, M.D., Ph.D., graduated from Harvard College and received an M.D. and a Ph.D. (the latter in pharmacology) from Harvard University and the Harvard–Massachusetts Institute of Technology (MIT) Program in Health Sciences and Technology. After completing residency and fellowship training in neurology at Harvard, he joined the faculty at Stanford University and began research into the mechanisms underlying pathological neuronal death. In 1991 he joined Washington University Medical School as head of the Neurology Department; there he also established the Center for the Study of Nervous System Injury, and directed the McDonnell Center for Cellular and Molecular Neurobiology. From 2001 until 2006, he was executive vice president for neuroscience at Merck Research Labs. Dr. Choi is currently executive director of the Comprehensive Neurosciences Initiative at Emory University. He is a fellow of AAAS, and a member of the IOM, the Executive Committee of the Dana Alliance for Brain Research Initiative, and the College of Physicians of Philadelphia. He has served as president of the Society for Neuroscience, vice president of the American Neurological Association, and chair of the U.S./Canada Regional Committee of the International Brain Research Organization. He has also served on the National Academy of Sciences' Board on Life Sciences, and Councils for the National Institute of Neurological Disorders and Stroke, the Society for Neuroscience, the Winter Conference for Brain Research, the International Society for Cerebral Blood Flow and Metabolism, and the Neurotrauma Society. He has been a member of advisory boards for the Christopher Reeve Paralysis Foundation, the Hereditary Disease Foundation, the Harvard–MIT Program in Health Sciences and Technology, the Queen's Neuroscience Institute in Honolulu, and the U.S. Food and Drug Administration (FDA), as well as for several university-based research consortia, biotechnology companies, and pharmaceutical companies.

Timothy Coetzee, Ph.D., is the National Multiple Sclerosis Society's vice president for discovery partnerships. In this capacity, Dr. Coetzee is responsible for the Society's strategic funding of biotechnology and pharmaceutical companies as well as partnerships with the financial and business communities. Dr. Coetzee received his Ph.D. in molecular biology from Albany Medical College and has been involved with multiple sclerosis (MS) research since then. He was a research fellow in the laboratory of society grantee Dr. Brian Popko at the University of North Carolina–Chapel Hill, and was the recipient of one of the society's Advanced Postdoctoral Fellowship Awards. After completing his training with Dr. Popko, Dr. Coetzee joined the faculty of the Department of Neuroscience at the University of Connecticut School of Medicine, where he conducted research that applied new technologies to understand how myelin is formed in the nervous system. He is the author of a number of research publications on the structure and function of myelin. Dr. Coetzee joined the society Home Office staff in fall 2000.

David H. Cohen, Ph.D., is a professor of psychiatry and biological sciences at Columbia University, where served as vice president and dean of the faculty of Arts and Sciences from 1995 to 2003. Prior to joining Columbia, he served as vice president for research and dean of the graduate school and subsequently as provost at Northwestern University. He has held professorships in physiology and/or neuroscience at Northwestern, State University of New York (SUNY)–Stony Brook, University of Virginia School of Medicine, and Case Western University School of Medicine. Dr. Cohen has held various elected offices in national and international organizations, including president of the Society for Neuroscience and chair of the Association of American Medical Colleges. He has served on various boards, including Argonne National Laboratory, the Fermi National Accelerator Laboratory, Zenith Electronics, and Columbia University Press. He has also served on numerous advisory committees for various organizations, including NIH, NSF, Department of Defense, and National Academies. Dr. Cohen received his B.A. from Harvard University and Ph.D. from the University of California–Berkeley, and was an NSF Postdoctoral Fellow at University of California–Los Angeles (UCLA).

Richard Frank, M.D., Ph.D., is the Vice President of Medical and Clinical Strategy for GE Healthcare. He has two decades of experience designing and implementing clinical trials in the pharmaceutical industry, and built the Experimental Medicine Department at Pharmacia before joining GE Healthcare in 2005. He is a past president and founding director of the Society of Non-invasive Imaging in Drug Development and a Fellow of the Faculty of Pharmaceutical Medicine, Royal College of Physicians. He serves on the scientific review board for the Institute for the Study of Aging and is a member of the editorial board of *Molecular Imaging and Biology*. Dr. Frank earned M.D. and Ph.D. (pharmacology) degrees concurrently and joined the pharmaceutical industry upon completion of his clinical training in 1985.

Richard Hodes, M.D., is the director of the National Institute of Aging at NIH. He is a diplomate of the American Board of Internal Medicine. In 1995 Dr. Hodes was elected as a member of the Dana Alliance for Brain Initiatives; in 1997 he was elected as a fellow of AAAS; and in 1999 he was elected to membership in the IOM. He also maintains an active involvement in research at NIH through his direction of the Immune Regulation Section, a laboratory devoted to studying regulation of the immune system, focused on cellular and molecular events that activate the immune response. In the past Dr. Hodes acted as a clinical investigator at the National Cancer Institute, then as the deputy chief and acting chief of the Cancer Institute's Immunology Branch. Since 1982 he has served as program coordinator for the U.S.–Japan Cooperative Cancer Research Program, and since 1992 on the scientific advisory board of the Cancer Research Institute. Dr. Hodes received his M.D. from Harvard Medical School. He completed a research fellowship at the Karolinska Institute in Stockholm and clinical training in internal medicine at Massachusetts General Hospital.

Steve Hyman, M.D., biography in Workshop Planning Committee.

Judy Illes, Ph.D., biography in Workshop Planning Committee.

Thomas R. Insel, M.D., biography in Workshop Planning Committee.

Story C. Landis, Ph.D., has been director of the National Institute for Neurological Disorders and Stroke (NINDS) since 2003. Dr. Landis oversees an annual budget of $1.5 billion and a staff of more than 900 scientists, physician-scientists, and administrators. The institute supports research by investigators in public and private institutions across the country, as well as by scientists working in its intramural laboratories and branches in Bethesda, MD. Since 1950, the institute has been at the forefront of U.S. efforts in brain research. Dr. Landis joined NINDS in 1995 as scientific director and worked with then-institute director Zach W. Hall, Ph.D., to coordinate and reengineer the Institute's intramural research programs. Between 1999 and 2000, under the leadership of NINDS director Gerald D. Fischbach, M.D., she led the movement, together with NIMH scientific director Robert Desimone, Ph.D., to bring a sense of unity and common purpose to 200 laboratories from 11 different NIH Institutes, all of which conduct leading-edge clinical and basic neuroscience research. Dr. Landis received her undergraduate degree in biology from Wellesley College and her master's and Ph.D. from Harvard University. After postdoctoral work at Harvard University studying transmitter plasticity in sympathetic neurons, she served on the faculty of the Harvard Medical School Department of Neurobiology. In 1985 she joined the faculty of Case Western Reserve University School of Medicine, where she held many academic positions, including chair of the Department of Neurosciences, which she was instrumental in establishing. Dr. Landis has made many fundamental contributions to the understanding of developmental interactions required for synapse formation. She has garnered many honors and awards and is an elected fellow of the Academy of Arts and Sciences, AAAS, and the American Neurological Association.

Ting Kai (TK) Li, M.D., earned his undergraduate degree from Northwestern University and his M.D. from Harvard University, and completed his residency training at Peter Bent Brigham Hospital in Boston, where he was named chief medical resident. He also conducted research at the Nobel Medical Research and Karolinska Institutes in Stockholm and served as deputy director of the Department of Biochemistry within the Walter Reed Army Institute of Research. Dr. Li joined the faculty at Indiana University as professor of medicine and biochemistry in 1971. Subsequently he was named the school's John B. Hickam Professor of Medicine and Professor of Biochemistry and later distinguished professor of medicine. In 1985 he became director of the Indiana Alcohol Research Center at the Indiana University School of Medicine, where he was also associate dean for research. Dr. Li is the recipient of numerous prestigious awards for his scientific accomplishments, including the Jellinek Award, the James B. Isaacson Award for Research in Chemical Dependency Diseases, and the R. Brinkley Smithers Distinguished Science Award. Dr. Li has also served in many prominent leadership and advisory positions, including past president of the Research Society on Alcoholism and as a member of the National Advisory Council on Alcohol Abuse and Alcoholism and the Advisory Committee to the Director, NIH. Dr. Li was elected to membership in the IOM in 1999 and is an honorary fellow of the United Kingdom's Society for the Study of Addiction.

Michael D. Oberdorfer, Ph.D., is the director of the Strabismus, Amblyopia and Visual Processing, and Low Vision and Blindness Rehabilitation Programs at the National Eye Institute of NIH. He is involved in a number of trans-NIH initiatives and activities in neuroscience and other areas, including the Coordinating Committee of the NIH Blueprint for Neuroscience Research. Before coming to NIH, he was a program officer at NSF, where he was involved in a number of activities, including directing the Developmental Neuroscience Program. Prior to that he was on the faculty of the University of Texas Medical School in Houston. He received his B.A. at Rockford College and his Ph.D. in zoology and neuroscience at the University of Wisconsin–Madison.

Kathie L. Olsen, Ph.D., became deputy director of NSF in 2005. She joined NSF from the Office of Science and Technology Policy (OSTP) in the Executive Office of the President, where she was the associate director and deputy director for science. Prior to the OSTP post, she served as the chief scientist at the National Aeronautics and Space Administration (NASA) and as the acting associate administrator for the new Enterprise in Biological and Physical Research. Before joining NASA in 1999, Dr. Olsen was the senior staff associate for the Science and Technology Centers in the NSF Office of Integrative Activities. From 1996 to 1997, she was a Brookings Institute legislative fellow and then an NSF detail in the Office of Senator Conrad Burns of Montana. Before her work on Capitol Hill, she served for 2 years as acting deputy director for the Division of Integrative Biology and Neuroscience at NSF. Dr. Olsen received her B.S. from Chatham College, majoring in both biology and psychology. She earned her Ph.D. in neuroscience at the University of California–Irvine. She was a postdoctoral fellow in the Department of Neuroscience at Children's Hospital of Harvard Medical School. Subsequently at SUNY–Stony Brook, she was both a research scientist at Long Island Research Institute and assistant professor in the Department of Psychiatry and Behavioral Science at the Medical School. Her research on neural and genetic mechanisms underlying development and expression of behavior was supported by NIH. Her awards include the NSF Director's Superior Accomplishment Award; the International Behavioral Neuroscience Society Award; the Society for Behavioral Neuroendocrinology Award for outstanding contributions in research and education; the Barnard Medal of Distinction, the college's most significant recognition of individuals for demonstrated excellence in conduct of their lives and careers; and NASA's Outstanding Leadership Medal.

Atul C. Pande, M.D., is senior vice president, Neurosciences MDC at GlaxoSmithKline. Previously he was the chief medical officer for Cenerx Biopharma. He has also served as vice president, GPM as well as vice president, neurosciences, for Pfizer Inc. Dr. Pande has extensive IND, NDA, and MAA experience in the areas of anxiety, depression, epilepsy, neuropathic pain, schizophrenia, traumatic brain injury, and Alzheimer's and Parkinson's diseases.

Steven Marc Paul, M.D., has been executive vice president of science and technology and president of the Lilly Research Laboratories (LRL) of Eli Lilly and Company since 2003. Dr. Paul joined Lilly in 1993 as a vice president of LRL responsible for central nervous system discovery and decision phase medical research. In 1996, Dr. Paul was appointed vice president (and in 1998 group vice president) of therapeutic area discovery research and clinical investigation. In this position his responsibilities included all therapeutic area discovery research, medicinal chemistry, toxicology/drug disposition, and decision phase (phase I/II) medical research. He and his leadership team were responsible for meeting the pipeline performance objectives of LRL and improving research and development (R&D) productivity, especially in discovery and the early phases of clinical development. In 2005, Dr. Paul was named Chief Scientific Officer of the Year as one of the Annual Pharmaceutical Achievement Awards. Prior to assuming his position at Lilly, Dr. Paul served as scientific director of NIMH. Dr. Paul received his B.A. in biology and psychology from Tulane University. He received his M.S. in anatomy (neuroanatomy) and his M.D. from Tulane University School of Medicine. Following an internship in neurology at Charity Hospital in New Orleans, he served as a resident in psychiatry and an instructor in the Department of Psychiatry at the University of Chicago, Pritzker School of Medicine. Dr. Paul also served as medical director in the Commissioned Corps of PHS, and maintained a private practice in psychiatry and psychopharmacology. He is board certified by the American

Board of Psychiatry and Neurology and has been elected a fellow in the American College of Neuropsychopharmacology (ACNP), served on the ACNP Council, and was elected president of ACNP (1999). He also serves on the executive board of the Pharmaceutical Research and Manufacturers of America's Science and Regulatory Committee and is incoming chairperson. Dr. Paul was appointed by the secretary of the U.S. Department of Health and Human Services (DHHS) to serve as a member of the Advisory Committee to the Director of NIH (2001–2006).

William Z. Potter, M.D., Ph.D., is vice president, Franchise Integrator Neuroscience, at Merck Research Laboratories. Prior to joining Merck, he served as the executive director and Lilly Clinical Research Fellow of the Neuroscience Therapeutic Area at Lilly Research Laboratories. He developed a Lilly/IU fellowship early in 1996 and was named professor of psychiatry at IUMC. Before being associated with Lilly Research Laboratories, he held the position of chief, Section on Clinical Pharmacology, Intramural Research Program at NIMH. He had been with PHS and NIH since 1971. He has authored more than 200 publications in the field of preclinical and clinical pharmacology, mostly focused on drugs used in affective illnesses and methods for evaluating drug effects in humans. He has received many honors during his career. Some of those include the 1975–1977 Falk Fellow, American Psychiatric Association; 1986 Meritorious Service Medal, PHS; and 1990 St. Elizabeth's Residency Program Alumnus of the Year Award.

Paul A. Sieving, M.D., Ph.D., became director of the National Eye Institute, NIH, in 2001. He came from the University of Michigan Medical School, where he was the Paul R. Lichter Professor of Ophthalmic Genetics and was the founding director of the Center for Retinal and Macular Degeneration in the Department of Ophthalmology and Visual Sciences. Dr. Sieving served as vice chair for clinical research for the Foundation Fighting Blindness from 1996 to 2001. He is on the Bressler Vision Award Committee and serves on the jury for the annual $1 million Award for Vision Research of the Champalimaud Foundation, Portugal. He was elected to membership in the American Ophthalmological Society in 1993 and the Academia Ophthalmologica Internationalis in 2005. He received an honorary Doctor of Science from Valparaiso University in 2003 and was named as one of the Best Doctors in America in 1998, 2001, and 2005. Dr. Sieving has received a number of awards, including the RPB Senior Scientific Investigator Award, 1998; the Alcon

Award, Alcon Research Institute, 2000; and the 2005 Pisart Vision Award from the New York Lighthouse International for the Blind. In 2006, Dr. Sieving was elected to the IOM.

Rae Silver, Ph.D., is the Helene L. and Mark N. Kaplan Professor of Natural and Physical Sciences and holds joint appointments at Barnard College and Columbia University. Dr. Silver is a fellow of the American Academy of Arts and Sciences, American Association of Arts and Sciences. She has participated extensively in scientific and educational activities, including serving as chair for NASA's Research Maximization and Prioritization Committee reviewing Scientific Priorities for the International Space Station; Society for Neuroscience Program committee (Theme E: Autonomic and Limbic System); chair, External Advisory Committee, NSF Center for the Study of Biological Rhythms at the University of Virginia; search committee for editor of journals, department chairs, and provost at various institutions; and panel member of a number of committees. As senior advisor at the National Science Foundation, she worked with NSF staffers in all the scientific directorates to create a series of workshops to examine opportunities for the next decade in making advances in neuroscience through the joint efforts of biologists, chemists, educators, mathematicians, physicists, psychologists, and statisticians. Dr. Silver's studies of the biological clock in the suprachiasmatic nucleus of the brain were the first to conclusively demonstrate that this brain tissue can be readily transplanted and restore function at a very high success rate in an animal model. The laboratory is renowned for analysis of the input, output, and intraneuronal circuits underlying the function of the brain's master clock. A second line of research entails the study of mast cells (renowned for their role in producing allergic reactions) in modulating brain function and as a major source of brain histamine. The research has been supported without interruption by NIH and NSF and others. Dr. Silver is deeply committed to educating undergraduate and graduate students, both at the national and institutional levels and in the hands-on context of the laboratory.

William H. Thies, Ph.D., is vice president for medical and scientific relations at the Alzheimer's Association, where he oversees the world's largest private, nonprofit Alzheimer's disease research grants program. Under his direction, the organization's annual grant budget has doubled, and the program has designated special focus areas targeting the relationships among cardiovascular risk factors and Alzheimer's disease, care-giving and care systems, and research involving diverse populations. He played a key role in launching *Alzheimer's & Dementia: The Journal of the Alzheimer's Association*, and in establishing the Research Roundtable, a consortium of senior scientists from industry, academia, and government who convene regularly to explore common barriers to drug discovery. In previous work at the American Heart Association (AHA) from 1988 to 1998, Dr. Thies formed a new stroke division that recently became the American Stroke Association. He also built the Emergency Cardiac Care Program, a continuing medical education program that trains more than 3 million professionals annually. He has worked with NINDS to form the Brain Attack Coalition. Prior to joining AHA, he held faculty positions at Indiana University in Bloomington and the University of Pittsburgh. Dr. Thies earned a B.A. in biology from Lake Forest College, and a Ph.D. in pharmacology from the University of Pittsburgh School of Medicine.

Roy E. Twyman, M.D., is vice president, Franchise Development, in the Central Nervous System/Pain Area of Johnson and Johnson Pharmaceutical Research and Development. In this position, he oversees licensing and acquisition efforts for neurology, psychiatry, and pain franchises while coordinating strategic activities for CNS discovery optimization, early human studies and proof of concept, new technologies, and cross-company projects. Additional oversight includes the pharmacogenomics and neuroimaging teams that support broad-based pharma R&D across all therapeutic areas. Before his work at Johnson and Johnson, Dr. Twyman was on faculty at the University of Utah and the University of Michigan. Dr. Twyman received his B.S. from Purdue University in Electrical Engineering. He earned his M.D. from the University of Kentucky and completed a neurology residency at University of Michigan.

Nora D. Volkow, M.D., became director of the National Institute on Drug Abuse in 2003. Dr. Volkow came to NIDA from Brookhaven National Laboratory (BNL), where she held concurrent positions, including associate director for life sciences, director of nuclear medicine, and director of the NIDA–Department of Energy Regional Neuroimaging Center. In addition, Dr. Volkow was a professor in the department of psychiatry and associate dean of the medical school at SUNY–Stony Brook. Dr. Volkow brings to NIDA a long record of accomplishments in drug addiction research. She is a recognized expert on the brain's dopamine system, with research focusing on the brains of addicted, obese, and aging individuals. Her studies have documented changes in the dopamine system affecting the actions of frontal brain regions involved with motivation, drive, and pleasure and the decline of brain dopamine function with age. Her work includes more than 350 peer-reviewed publications, 3 edited books, and more than 50 book chapters and non-peer reviewed manuscripts. The recipient of multiple awards, she was elected to membership in the IOM and was named "Innovator of the Year" in 2000 by *U.S. News and World Report*. Dr. Volkow received her B.A. from Modern American School, Mexico City, Mexico; her M.D. from the National University of Mexico, Mexico City; and her postdoctoral training in psychiatry at New York University. In addition to BNL and SUNY–Stony Brook, Dr. Volkow has worked at the University of Texas Medical School and Sainte Anne Psychiatric Hospital in Paris, France.

Christian G. Zimmerman, M.D., FACS, M.B.A., biography in Workshop Planning Committee.

Stevin H. Zorn, Ph.D., is vice president and head of Central Nervous System Disorders Research at Pfizer Global Research and Development, and also coleads Pfizer's CNS Therapeutic Area Leadership Team. He received a B.S. in chemistry from Lafayette College, and an M.S. and a Ph.D. in biomedical sciences with an emphasis on toxicology and neuropharmacology, respectively. Dr. Zorn conducted postdoctoral research studies in Paul Greengard's Laboratory of Molecular and Cellular Neuroscience at Rockefeller University before joining Pfizer in 1989. Dr. Zorn has coauthored numerous scientific research communications and patents and has contributed to the advancement of a wide variety of drug candidates, some of which are now helping to improve the lives of patients suffering from CNS-related illness.

INVITED SPEAKERS

Arthur Beaudet, M.D., received a B.S. in biology from College of the Holy Cross and an M.D. from Yale. He then did 2 years of pediatrics residency at Johns Hopkins and spent 2 years as a research associate at NIH before going to Baylor College of Medicine in 1971, where he remains. Dr. Beaudet has published more than 200 original research articles in diverse aspects of mammalian genetics. His contributions included the demonstration of mutations in cultured somatic cells in the

1970s, a time when such evidence was still considered novel. He published extensively on inborn errors of metabolism, particularly on urea cycle disorders. His group was the first to describe uniparental disomy in humans in 1988. He has longstanding interests in somatic gene therapy and in cystic fibrosis. More recently his major focus has been on genomic imprinting as it relates to Prader-Willi and Angelman syndromes, including identification of the gene causing Angelman syndrome. Dr. Beaudet is well known as one of the editors of the *Metabolic and Molecular Bases of Inherited Disease* tome for the 6th through 8th editions, and he has served on many editorial boards and national review panels. He was president of the American Society of Human Genetics in 1998 and is an elected member of the Association of American Physicians and the IOM. Dr. Beaudet is currently the Henry and Emma Meyer Distinguished Service Professor and chair in the Department of Molecular and Human Genetics at Baylor College of Medicine in Houston.

Sallie Bernard is a cofounder and the executive director of SafeMinds. She serves as the chair of the board of directors of Cure Autism Now, one of the largest funders of biomedical research for autism. She was formerly executive director of the New Jersey Chapter of Cure Autism Now, helping to secure millions of dollars in funding from the state of New Jersey for autism research and treatment. She was also a member of the Founders Forum for the Autism Center at UMDNJ in New Jersey. Ms. Bernard has testified before Congress as well as made a presentation to the IOM. She has published a number of research papers and letters in science journals, and participates in several government committees addressing the effect of mercury on neurodevelopment. Ms. Bernard is a cofounder and president of Extreme Sports Camp, a nonprofit summer camp for older children and teenagers with autism. She is the founder and former president of ARC Research, a full-service market research and marketing consulting firm that she sold in 2004. She graduated from Radcliffe College, Harvard University. One of her children has autism.

Laura Bono, biography in Workshop Planning Committee.

Henry Falk, M.D., serves as director, Coordinating Center for Environmental Health and Injury Prevention (CCEHIP), one of four Coordinating Centers at CDC. CCEHIP includes the National Center for Environmental Health/Agency for Toxic Substances and Disease Registry (NCEH/ATSDR) and the National Center for Injury Prevention and

Control. Prior to this, he served as director for both NCEH and ATSDR. Dr. Falk is also a member of the Executive Leadership Board of CDC, where he arrived in 1972. He is a 30-year veteran of the PHS Commissioned Corps. This service culminated with his being named Rear Admiral and Assistant U.S. Surgeon General. Dr. Falk earned his M.D. from the Albert Einstein College of Medicine. He received a master's degree from Harvard School of Public Health, and he is board certified in pediatrics as well as public health and general preventive medicine. His honors include the Vernon Houk Award for Leadership in Preventing Childhood Lead Poisoning and the Homer C. Calver Award in environmental health from the American Public Health Association. He has also received CDC's William C. Watson Jr. Medal of Excellence, as well as PHS's Distinguished Service Award.

Gary W. Goldstein, M.D., is chair of the Autism Speaks scientific affairs committee, and president and CEO of the Kennedy Krieger Institute, one of the nation's leading treatment centers for autism and other developmental disorders. He is also a professor of neurology and pediatrics at Johns Hopkins University's School of Medicine and a professor of environmental health sciences at the University's School of Hygiene and Public Health. One of the leading researchers of neurological functions and defects, Dr. Goldstein has helped gain international recognition for the Kennedy Krieger Institute through his studies of children with a wide range of disabilities, from rare genetic disorders to common learning problems. More than 10,000 children with disabilities visit the Kennedy Krieger Institute every year.

Martha Herbert, M.D., Ph.D., is an assistant professor of neurology at Harvard Medical School, a pediatric neurologist at Massachusetts General Hospital and at the Center for Child and Adolescent Development of Cambridge Health Alliance, a member of the MGH Center for Morphometric Analysis, and an affiliate of the Harvard–MIT–MGH Martinos Center for Biomedical Imaging. She earned her medical degree at the Columbia University College of Physicians and Surgeons. Prior to her medical training, she obtained a doctoral degree at the University of California–Santa Cruz, studying evolution and development of learning processes in biology and culture in the History of Consciousness program, and then did postdoctoral work in the philosophy and history of science. She trained in pediatrics at Cornell University Medical Center, and in neurology and child neurology at MGH, where she has remained.

She received the first Cure Autism Now Innovator Award; she is the co-chair of the Environmental Health Advisory Board of the Autism Society of America. Her research program utilizes multimodal brain imaging techniques, including MRI, EEG, and MEG, in coordination with clinical observation, metabolic biomarkers, and animal studies, to study the physiological underpinnings of autism, aiming toward understanding what makes some autistic brains unusually large, what causes altered brain connectivity, how we can develop measures sensitive to changes in brain function that could result from treatment interventions, and what might be potential domains of plasticity and targets for intervention.

Irva Hertz-Picciotto, Ph.D., M.P.H., received her B.A. in mathematics, M.A. in biostatistics, and Ph.D./M.P.H. in epidemiology from the University of California–Berkeley. After 12 years on the faculty at University of North Carolina (UNC)–Chapel Hill, she returned to California to join the University of California–Davis Department of Public Health Sciences (formerly the Department of Epidemiology and Preventive Medicine). Her research interests are in environmental exposures (metals, pesticides, PCBs, air pollution), pregnancy outcomes (spontaneous abortion, fetal growth, early child development), and epidemiologic methods (left truncation in survival analysis, the "healthy worker survivor bias," timing issues, and use of epidemiologic data in quantitative risk assessment). She authored the chapter "Environmental Epidemiology" in the textbook *Modern Epidemiology* (Rothman and Greenland, and currently serves on editorial boards for the *American Journal of Epidemiology*, *Environmental Health Perspectives*, and *Epidemiology*, as well as on scientific advisory boards for the U.S. Environmental Protection Agency (EPA) and the National Institute for Occupational Safety and Health. Previously she served on the Governor's Carcinogen Identification Committee for the state of California, the Board of Scientific Counselors of the National Toxicology Program, and the Scientific Advisory Panel for the Interagency Coordinating Committee on Autism Research. Dr. Hertz-Picciotto chaired the IOM Committee on the Health Effects in Vietnam Veterans of Exposure to Agent Orange and Other Herbicides in 2000 and 2002. She directed the program in reproductive epidemiology at UNC–Chapel Hill and is the deputy director of the Center for Children's Environmental Health at UC–Davis, focused on autism and other neurodevelopmental disorders.

Thomas R. Insel, M.D., biography in Workshop Planning Committee.

S. Jill James, Ph.D., is a research biochemist with more than 25 years of experience studying metabolic biomarkers of disease susceptibility. She received her B.S. in biology from Mills College in Oakland, CA, and her Ph.D. in nutritional biochemistry from UCLA. She is a professor in the Department of Pediatrics at the University of Arkansas for Medical Sciences and director of the Autism Metabolic Genomics Laboratory at the Arkansas Children's Hospital Research Institute. Before transferring to the University, she was a senior research scientist at the FDA National Center for Toxicological Research, where she directed a laboratory focused on DNA methylation and cancer susceptibility. Her research career has been focused on defining gene–environment interactions that increase susceptibility to cancer, Down syndrome, and most recently, autism. She has published more than 120 peer-reviewed papers and recently received the American Society for Nutritional Sciences award for innovative research in the understanding of human nutrition.

Philip J. Landrigan, M.D., is a pediatrician, epidemiologist, and internationally recognized leader in public health and preventive medicine. He has been a member of the faculty of Mount Sinai School of Medicine since 1985 and chair of the Department of Community and Preventive Medicine since 1990. Dr. Landrigan graduated from Harvard Medical School. In 1977, he received a Diploma of Industrial Health from the London School of Hygiene and Tropical Medicine. He completed a residency in pediatrics at Boston Children's Hospital. He then served for 15 years as an epidemic intelligence service officer and medical epidemiologist at CDC and the National Institute for Occupational Safety and Health. In 1987, Dr. Landrigan was elected as a member of the IOM. He is editor-in-chief of the *American Journal of Industrial Medicine* and previously was editor of *Environmental Research.* He has chaired committees at the National Academy of Sciences on *Environmental Neurotoxicology* and on *Pesticides in the Diets of Infants and Children.* The NAS report that he directed on pesticides and children's health was instrumental in securing passage of the Food Quality Protection Act, the major federal pesticide law in the United States. From 1995 to 1997, Dr. Landrigan served on the Presidential Advisory Committee on Gulf War Veteran's Illnesses. In 1997–1998, he served as senior advisor on children's health to the EPA administrator and was instrumental in helping to establish a new Office of Children's Health Protection at EPA. From 2000 to 2002, Dr. Landrigan served on the Armed Forces Epidemiological Board. He served from 1996 to 2005 in the Medical Corps of the U.S.

Naval Reserve. He continues to serve as deputy command surgeon general of the New York Naval Militia. Dr. Landrigan is known for his many decades of work in protecting children against environmental threats to health, most notably lead and pesticides. He has been a leader in developing the National Children's Study, the largest study of children's health and the environment ever launched in the United States. He has been centrally involved in the medical and epidemiologic studies that followed the destruction of the World Trade Center on September 11, 2001.

Pat Levitt, Ph.D., received his Ph.D. in neuroscience at the University of California–San Diego. He completed a postdoctoral fellowship in neuroscience at Yale University School of Medicine. He was named a McKnight Foundation Scholar in 2002. Dr. Levitt also is an elected fellow of AAAS and chair of the Scientific Advisory Board of Cure Autism Now. Dr. Levitt is a member of the Dana Alliance for Brain Initiatives, the National Scientific Council on the Developing Child, and the National Advisory Mental Health Council for NIMH. Dr. Levitt's research interests are in the development of brain circuits that control learning and emotion. His clinical genetics and basic research studies focus on understanding the basis of neurodevelopmental and neuropsychiatric disorders, and how genes and the environment together influence typical and atypical development. He has received a number of research grants from NIH, the McKnight Endowment Fund, the Joseph and Esther Klingenstein Foundation, the March of Dimes, and other foundations. Dr. Levitt serves on the editorial boards of *Biological Psychiatry*, *Cerebral Cortex*, and *Neuron,* and he was senior editor for the *Journal of Neuroscience.* He is the author or coauthor of more than170 scientific papers. Dr. Levitt is a frequently invited speaker at national and international seminars and conferences, as well as public education and policy forums that promote the health and education of children.

Ian Lipkin, **M.D.,** is professor of epidemiology in the Mailman School of Public Health, and director of the Columbia Center for Infection and Immunity. Through June 2002 Dr. Lipkin also held academic positions at the University of California–Irvine. He is internationally recognized as an authority on the use of molecular biological methods for pathogen discovery and the role of immune and microbial factors in neurologic and neuropsychiatric diseases. Dr. Lipkin received a B.A. from Sarah Lawrence College, where he studied cultural anthropology, philosophy, and literature, and an M.D. from Rush Medical College. His postgraduate training included clerkship at the Queen Square Institute of Neurology in London; internship in medicine at the University of Pittsburgh; residency in internal medicine at the University of Washington; residency in neurology at UCSF; and fellowship in neurovirology and molecular neurobiology at the Scripps Research Institute in La Jolla, CA. His honors include National Multiple Sclerosis Society Postdoctoral Fellowship; Clinical Investigator Development Award, NIH, National Institute of Neurological and Communicative Disorders and Stroke; Pew Scholar; Louise Turner Arnold Chair in the Neurosciences; and Ellison Medical Foundation Senior Scholar in Global Infectious Diseases.

Fernando D. Martinez, M.D., is director of the Arizona Respiratory Center and Swift-McNear Professor of Pediatrics at the University of Arizona in Tucson. His major research interests include the natural history of childhood asthma, the genetic epidemiology of asthma and related conditions, and the early development of the immune system as a risk factor for the development of asthma. Dr. Martinez is the director of one of five centers participating in the Childhood Asthma Research and Education Network, a national effort funded by the National Heart, Lung, and Blood Institute. He is also the recipient of two other current NIH grants. Dr. Martinez is an associate editor of *Thorax* and is a reviewer for various journals, including *Lancet*, *New England Journal of Medicine*, *Journal of the American Medical Association*, and *European Respiratory Journal*. He has written more than 150 journal articles, book chapters, editorials, and abstracts, and he has been an invited lecturer at numerous national and international conferences. Dr. Martinez received a medical license (equivalent to an M.D.) from the University of Chile in Santiago. He then completed a medical degree and a fellowship in pediatrics with a specialization in pulmonology at the University of Rome in Rome, Italy.

Larry L. Needham, Ph.D., is chief of the Organic Analytical Toxicology Branch of the National Center for Environmental Health, CDC. He has served at CDC for more than 30 years in the area of assessing human exposure to environmental chemicals through biomonitoring. Dr. Needham has authored or coauthored about 400 publications in this area, with special emphasis on polychlorinated dibenzo-*p*-dioxins, furans, and biphenyls; pesticides; phthalates; perfluorinated chemicals; volatile organic chemicals; and inorganic elements. Dr. Needham has received many awards, including PHS's Special Recognition and Superior Service Award; CDC's honor award for outstanding scientific leadership; and in 2006 the International Society of Exposure Analysis's (ISEA's) most prestigious award, the Wesolowski Award, for his biomonitoring work. Dr. Needham serves on advisory boards for many scientific organizations and studies. In addition, he is a past president of ISEA, editor of *Chemosphere: Dioxins and Persistent Organic Pollutants,* and federal co-chair of the exposure workgroup for planning for the National Children's Study. He is also the initial recipient of ISEA's Distinguished Lecturer Award.

Craig Newschaffer, Ph.D., is professor and chair of the Department of Epidemiology and Biostatistics at the Drexel University School of Public Health. Dr. Newschaffer recently joined the Drexel faculty, coming from the Department of Epidemiology at the Johns Hopkins Bloomberg School of Public Health. At Johns Hopkins, Dr. Newschaffer founded and directed the Center for Autism and Developmental Disabilities Epidemiology, one of five federally funded centers of excellence in autism epidemiology. Major initiatives included the development of methods for monitoring autism spectrum disorders prevalence and participation in the largest population-based epidemiologic study of autism risk factors to date: the National Centers for Autism and Developmental Disabilities Research and Epidemiology (CADDRE) Study of Autism and Child Development. Dr. Newschaffer also is engaged in other projects focusing on how particular genes might interact with environmental exposures to increase autism risk. He recently began a collaboration with Peking University to explore approaches for conducting epidemiologic research on autism in China. Dr. Newschaffer is an associate editor of the *American Journal of Epidemiology* and a member of the editorial board of the journal *Developmental Epidemiology*.

Mark Noble, Ph.D., is a pioneering researcher in the field of stem cell biology and CNS development. He was codiscoverer of the first progenitor cell to be isolated from the CNS, the progenitor cell that gives rise to myelin-forming oligodendrocytes. His laboratory then discovered cell–cell interactions and specific mitogens that control the division of these cells, along with conditions allowing greatly enhanced cell expansion in vitro. These discoveries led to the first use of purified precursor cell populations for repair of experimental CNS lesions. His laboratory also discovered adult-specific populations of progenitor cells, and the team of researchers with whom he works has played a central role in the discovery, isolation, and characterization of nearly all of the lineage-restricted progenitor cell populations that have been isolated from the developing CNS, characterized at the clonal level, and transplanted back into the CNS. Dr. Noble's current research is focused on developing a comprehensive approach to the field of stem cell medicine, research which includes topics such as identifying the optimal cells for enhancing repair of spinal cord injury; the central importance of precursor cell dysfunction in developmental maladies; and the discovery of molecular mechanisms that underlie effects of environmentally relevant levels of chemically diverse toxicants on CNS precursor cells and that integrate stem cell biology, redox biology, signaling pathway analysis, and toxicology into a mechanistic framework. Dr. Noble is professor of genetics, neurobiology, and anatomy at the University of Rochester Medical Center, and is codirector of the New York State Center of Research Excellence for Spinal Cord Injury.

Isaac Pessah, Ph.D., is professor and chair of the Department of Molecular Biosciences in the School of Veterinary Medicine at the University of California–Davis. He is also director of the NIEHS/EPA Children's Center for Environmental Health and Disease Prevention: Environmental Factors in the Etiology of Autism. Dr. Pessah is a toxicologist with research interest in the area of molecular and cellular mechanisms regulating signaling in excitable cells. His current research focuses on the structure, function, and pharmacology of the ryanodine-sensitive calcium channels (RyRs) found in sarcoplasmic and endoplasmic reticulum of muscle cells and neurons. His laboratory is actively studying how dysfunction of RyRs complexes contribute to genetic diseases and how genetic alteration of RyRs and environmental factors interact to influence neurodevelopment by utilizing cellular, biochemical, and molecular investigations of calcium-signaling pathways. He is a senior member of the NIEHS Center of Excellence in Toxicology and the Superfund Basic Research Program.

William F. Raub, Ph.D., is science advisor to the secretary of Health and Human Services and deputy assistant secretary for Public Health Emergency Preparedness. Dr. Raub was acting assistant secretary for Public Health Emergency Preparedness from 2003 to 2004, principal deputy assistant secretary for Planning and Evaluation from 2000 to 2002, acting assistant secretary for Planning and Evaluation during 2001 and again during 2003, and deputy assistant secretary for Science Policy from 1995 to 2000. He was the science advisor to the EPA administrator from 1992 to 1995 after a 1-year assignment as special assistant for Health Affairs in the Office of Science and Technology Policy, Executive Office of the President of the United States. Prior to that, he was the deputy director of NIH from 1986 through 1991. From 1989 through 1991, he was the acting director, NIH. From 1978 to 1986, Dr. Raub served first as associate director, and later deputy director, for Extramural Research and Training at NIH. He was associate director of the National Eye Institute from 1975 to 1978 and chief of the Biotechnology Resources Branch in the Division of Research Resources from 1969 to 1975. From 1966 through 1979, Dr. Raub led the development of the PROPHET system, the first integrated array of computer-based tools for the study of the relationships between molecular structures and biological effects. Dr. Raub has received numerous awards from external organizations for his government service, including the Society of Research Administrators' Award for Distinguished Contribution to Research Administration, the American Medical Association's Nathan Davis Award, and election as a fellow of the National Academy of Public Administration. In addition, within DHHS, he has twice been presented the Distinguished Service Award and has received the Presidential Meritorious Executive Rank Award and the Presidential Distinguished Rank Award. Dr. Raub earned an A.B. in biology from Wilkes College and a Ph.D. in physiology from the University of Pennsylvania, where he also was awarded an NSF graduate fellowship and was a fellow of the Pennsylvania Plan.

Lyn Redwood, R.N., M.S.N., CRNP, is a nurse practitioner and has worked in the nursing profession for 25 years specializing in pediatrics and women's health care. In the late 1990s, she became involved in autism research when her son was diagnosed with pervasive developmental disorder, not otherwise specified and found to be mercury toxic. Ms. Redwood is coauthor of *Autism: A Novel Form of Mercury Toxicity* and has testified before the Government Reform Committee on Mercury in Medicine on the question: Are we taking unnecessary risks? As a writer and researcher on autism and mercury toxicity, Ms. Redwood has been published in *Neurotoxicology*, *Medical Hypothesis*, *Molecular Psychiatry*, *Mothering Magazine*, and *Autism-Aspergers Digest*. She has also appeared on "Good Morning America" with Diane Sawyer and has been interviewed by *U.S. News and World Report*, *Wired Magazine*, and numerous other publications. Ms. Redwood is cofounder of the Coalition for SafeMinds and was featured prominently in the book "Evidence of Harm" by David Kirby.

Diana E. Schendel, Ph.D., is lead health scientist and epidemiology team lead in the Developmental Disabilities Branch, National Center on Birth Defects and Developmental Disabilities, CDC. She serves as science liaison for CDC's CADDRE and is principal investigator for CDC's Georgia CADDRE study site. She coordinates scientific activities in CADDRE, including the CADDRE multisite study of autism (Study to Explore Early Development, or SEED), the largest epidemiologic study of the causes of autism planned to date. She serves as science liaison and CDC principal investigator for CDC's Collaborative Public Health Research Program in Denmark with the Danish Agency for Science, Technology and Innovation. Her professional research interests are in developmental disabilities epidemiology. She has been recognized for her work in autism (Secretary's Award for Distinguished Service [2005], Autism Public Health Response Team, Secretary of Health and Human Services; CDC and ATSDR Group Honor Award [2002]), Research Operational, Autism Public Health Response Team) and cerebral palsy. She is a member of the epidemiology subcommittee of the Scientific Advisory Board of Autism Speaks and Scientific Advisory Board of the European Autism Information System. She received a B.S. in both biology and anthropology from Florida State University and an M.A. and a Ph.D. in anthropology from Pennsylvania State University. She began her career at Tufts University in the Department of Sociology and Anthropology, then joined CDC's Division of Birth Defects and Developmental Disabilities as an epidemiologist.

David A. Schwartz, M.D., biography in Workshop Planning Committee.

Theodore A. Slotkin, Ph.D., received a Ph.D. in pharmacology and toxicology from the University of Rochester. He has done extensive research in the areas of developmental pharmacology and toxicology, neuropharmacology and neurochemistry, and cell differentiation and growth regulation. His research is aimed toward understanding the interaction of drugs, hormones, and environmental factors with the developing organism, with particular emphasis on the fetal and neonatal nervous systems. His most notable achievements concern the effects of fetal exposure to drugs of abuse, especially tobacco and nicotine; drugs used in preterm labor; and neuroactive pesticides. He has received numerous honors and awards for his research work, notably the Alton Ochsner Award Relating Smoking and Health, the John J. Abel Award in Pharmacology, and the Otto Krayer Award in Pharmacology, and has published more than 480 peer-reviewed articles. He has served on NIH Consensus Panels on Pharmacotherapies for Smoking Cessation During Pregnancy and on The Use of Antenatal Steroids. He has chaired review boards for the California Tobacco-Related Diseases Research Program, and he serves on the editorial boards of three scholarly journals. He is among the 1 percent of "Most Cited Scientists in Pharmacology & Toxicology" identified by the Institute for Scientific Information.

Sarah Spence, M.D., Ph.D., is a board-certified pediatric neurologist with a doctorate in neuropsychology and clinical and research expertise in autism spectrum disorders. She received her Ph.D. in cognitive neuroscience from UCLA in 1992 and her M.D. from UCSF in 1995. She completed her medical training in pediatrics and neurology at UCLA in 2000 and a fellowship in neurobehavioral genetics in 2001 with Dr. Daniel Geschwind while working with the Autism Genetic Resource Exchange (AGRE), a gene bank created by the Cure Autism Now foundation. She then served on the UCLA medical school faculty, where she was a member of the Center for Autism Research and Treatment, responsible for overseeing research recruitment and assessment. She was medical director of the Autism Evaluation Clinic, with an active practice specializing in children with autism spectrum disorder. Dr. Spence was recently recruited to the Division of Intramural Research at NIMH, where she is contributing to the design and administration of various clinical research protocols examining the phenomenology of and novel treatments for children with autism spectrum disorders. She continues to work with community organizations as a neurological consultant to AGRE, a member of the Treatment Advisory Board and Autism Treatment Network steering committees for CAN, and the treatment subcommittee of the Scientific Advisory Committee for Autism Speaks. Her research interests include the role of epilepsy in autism, examination of the autism phenome, clinical trials in novel treatments, and the genetics of autism spectrum and related developmental disorders.

Ezra Susser, M.D., Ph.D., M.P.H., is the Anna Cheskis Gelman and Murray Charles Gelman Professor and chair of the Department of Epidemiology, and professor of psychiatry in the New York State Psychiatric Institute. His primary research has been on the epidemiology of mental disorders and on examining the role of early life experience in health and disease throughout the life course. His international collaborative birth cohort research program (The Imprints Center) seeks to uncover the causes of a broad range of disease and health outcomes, including psychiatric and neurodevelopmental disorders such as autism, schizophrenia, and attention deficit hyperactivity disorder, among others. Among the risk factors explored are prenatal exposures to infectious disease and toxic chemicals, childhood nutrition and environment, and genetics, as well as the interplay of genetic and environmental risk factors. Dr. Susser has also focused on public health initiatives regarding HIV/AIDS throughout his career, both locally and internationally.

Susan Swedo, M.D., biography in Workshop Planning Committee.

David R. Walt, Ph.D., is Robinson Professor of Chemistry at Tufts University and a Howard Hughes Medical Institute professor. He received a B.S. in chemistry from the University of Michigan and a Ph.D. in chemical biology from SUNY–Stony Brook. After postdoctoral studies at MIT, he joined the chemistry faculty at Tufts. He served as chemistry department chair from 1989 to 1996. Dr. Walt serves on many government advisory panels and boards and serves on the editorial advisory boards for numerous journals. From 1996 to 2003, he was executive editor of *Applied Biochemistry and Biotechnology*. Dr. Walt is the scientific founder and a director of Illumina, Inc. He has received numerous national and international awards and honors and is a fellow of AAAS. Dr. Walt has published over 200 papers, holds more than 40 patents, and has given hundreds of invited scientific presentations.

Allen J. Wilcox, M.D., Ph.D., is a senior investigator in the Epidemiology Branch of NIEHS, NIH, where he has worked since 1979. He was chief of the Epidemiology Branch from 1991 to 2001, and since 2001 has served as the editor-in-chief of the journal *Epidemiology*. He is past president of the American Epidemiological Society, the Society for Epidemiologic Research, and the Society for Pediatric Epidemiologic Research. He holds adjunct appointments as professor of epidemiology at the University of North Carolina and the University of Bergen (Norway), and has served on three IOM committees. He is a fellow in the American College of Epidemiology. His research area is reproductive and perinatal epidemiology, with special interest in early pregnancy, pregnancy loss, and fetal growth and development. His current research project is on the genetic and environmental causes of cleft lip and cleft palate. He received a B.A. in psychology and an M.D. from the University of Michigan, and an M.P.H. in maternal and child health and a Ph.D. in epidemiology from UNC–Chapel Hill.

STAFF

Bruce M. Altevogt, Ph.D., is a senior program officer in the Board on Health Sciences Policy at the IOM. His primary interests focus on policy issues related to basic research and preparedness for catastrophic events. He received his Ph.D. from Harvard University's Program in Neuroscience. Following over 10 years of research, Dr. Altevogt joined the National Academies as a science and technology policy fellow with the Christine Mirzayan Science & Technology Policy Graduate Fellowship Program. Since joining the Board on Health Sciences Policy, he has been a program officer on multiple IOM studies including, *Sleep Disorders and Sleep Deprivation: An Unmet Public Health Problem, The National Academies' Guidelines for Human Embryonic Stem Cell Research: 2007 Amendments,* and *Assessment of the NIOSH Head-and-Face Anthropometric Survey of U.S. Respirator Users*. He is currently serving as the director of the Neuroscience and Nervous System Disorders Forum and a co-study director on the National Academy of Sciences Human Embryonic Stem Cell Research Advisory Committee. He received his B.A. from the University of Virginia in Charlottesville, where he majored in biology and minored in South Asian studies.

Andrew Pope, Ph.D., is director of the Board on Health Sciences Policy at the IOM. With a Ph.D. in physiology and biochemistry, his primary interests focus on environmental and occupational influences on human health. Dr. Pope's previous research activities focused on the neuroendocrine and reproductive effects of various environmental substances on food-producing animals. During his tenure at the National Academies and since 1989 at the IOM, Dr. Pope has directed numerous studies; topics include injury control, disability prevention, biological markers, neurotoxicology, indoor allergens, and the enhancement of environmental and occupational health content in medical and nursing school curriculums. Most recently, Dr. Pope directed studies on NIH priority-setting processes, organ procurement and transplantation policy, and the role of science and technology in countering terrorism.

Sarah L. Hanson is a senior program associate in the Board on Health Sciences Policy at the IOM. Ms. Hanson previously worked for the Committee on Sleep Medicine and Research. She is currently the senior program associate for the Forum on Neuroscience and Nervous System Disorders. Prior to joining the IOM, she served as research and program assistant at the National Research Center for Women & Families. Ms. Hanson has a B.A. from the University of Kansas with a double major in political science and international studies. She is currently taking premedicine courses at the University of Maryland and hopes to attend medical school in the future.

Afrah J. Ali is a senior program assistant for the Board on Health Sciences Policy at the IOM. Earlier, she studied biology at Howard University. Ms. Ali has 7 years of integrated project management, executive administration, publishing, event planning, research, and marketing experience. Her previous positions include marketing specialist at Standard and Poor's E-marketing division in New York City.

Lora K. Taylor is a senior program assistant for the Board on Health Sciences Policy at the IOM. She has 15 years of experience working at the National Academies. Before joining the IOM she served as the administrative associate for the Report Review Committee and the Division on Life Sciences' Ocean Studies Board. Ms. Taylor has a B.A. from Georgetown University with a double major in psychology and fine arts.